DIGITAL THEORY
AND EXPERIMENTATION
USING INTEGRATED
CIRCUITS

PRENTICE-HALL SERIES IN ELECTRONIC TECHNOLOGY

Dr. Irving L. Kosow, *editor*
Charles M. Thomson and Joseph J. Gershon, *consulting editors*

DIGITAL THEORY AND EXPERIMENTATION USING INTEGRATED CIRCUITS

Morris E. Levine

Associate Professor
Staten Island Community College
City University of New York

PRENTICE-HALL, INC., ENGLEWOOD CLIFFS, NEW JERSEY

Library of Congress Cataloging in Publication Data

Levine, Morris E
    Digital theory and experimentation using
integrated circuits.

    (Prentice-Hall series in electronic technology)
    Bibliography: p.
    1. Digital electronics—Laboratory manuals.
2. Integrated circuits. I. Title.
TK7868.D5L48   621.381'73   73-12863
ISBN   0-13-212258-8

Printed in the United States of America

10 9 8 7 6 5 4 3 2 1

PRENTICE-HALL INTERNATIONAL, INC., London
PRENTICE-HALL OF AUSTRALIA, PTY. LTD., Sydney
PRENTICE-HALL OF CANADA, LTD., Toronto
PRENTICE-HALL OF INDIA PRIVATE LIMITED, New Delhi
PRENTICE-HALL OF JAPAN, INC., Tokyo

To
Pearl, Estelle, and Carl

CONTENTS

PREFACE ix

TO THE INSTRUCTOR xi

Experiment 1
BASIC LOGIC FUNCTIONS 1

Experiment 2
BOOLEAN ALGEBRA AND SIMPLIFICATION OF LOGIC EQUATIONS 11

Experiment 3
DE MORGAN'S THEOREM 23

Experiment 4
TTL NAND/NOR GATES—DEFINITIONS AND OPERATION 31

Experiment 5
THE "EXCLUSIVE OR" AND ITS APPLICATIONS 47

Experiment 6
FULL ADDER/FULL SUBTRACTOR 59

Experiment 7
BISTABLE OR FLIP–FLOP (FF) 71

Experiment 8
BINARY COUNTERS/THE BINARY NUMBER SYSTEM 95

Experiment 9
DIVIDE–BY–N COUNTERS; DECADE COUNTERS 109

Experiment 10
SHIFT REGISTERS AND RING COUNTERS 125

Experiment 11
PULSE FORMING 141

Experiment 12
PULSE SHAPING/SCHMITT TRIGGER 151

Experiment 13
DECODING-ENCODING 159

Experiment 14
RANDOM ACCESS (RAM)—SCRATCH PAD MEMORIES 171

Experiment 15
THE OPERATIONAL AMPLIFIER 181

Experiment 16
DIGITAL-TO-ANALOG (D/A) AND ANALOG-TO-DIGITAL (A/D) CONVERSION 195

Appendix A
CATHODE RAY OSCILLOSCOPE (CRO)* 205

Appendix B
LOGIC SYMBOLS 207

Appendix C
THE "UNIT LOAD" CONCEPT 209

Appendix D
IC LOGIC DIAGRAMS AND PIN CONNECTIONS 211

Appendix E
GLOSSARY OF LOGIC AND INTEGRATED CIRCUIT TERMINOLOGY 217

Appendix F
REFERENCES 223

Appendix G
IC SOCKET, SWITCH BANK, AND EQUIPMENT 225

PREFACE

This laboratory manual has been developed to provide laboratory training and experience in digital electronics using integrated circuits in the laboratory environment of the industrial electronics laboratory. As in the industrial electronics laboratory all of the measurements (with very minor exceptions) are made using the cathode ray oscilloscope. This manual therefore serves two purposes. Besides providing the digital techniques experience, it also provides intensive and thorough training in the use of the CRO as a universal tool for the measurement and analysis of digital circuits.

There are five primary techniques used for the analysis of digital circuits and theory. They are:

A. Voltage levels
B. Logic levels; 1 or 0, T or F, H or L, ON or OFF
C. Logic diagrams
D. Mapping methods
E. Sequential operations. The CRO is used to demonstrate and perform the analysis.

All five methods are integrated to advance understanding of digital principles.

Sixteen experiments are included in this manual to provide thorough coverage of digital principles. They begin with a series of experiments on the principles of logic. These are followed with experiments on arithmetic operations. Several experiments cover the material on counters, counter techniques, the binary number system, decoding-encoding, memory techniques, pulse generation and pulse shaping/IC Schmitt Trigger. The operational amplifier has become of great importance in digital techniques. Two experiments develop the principles and applications of the OP-AMP in D/A and A/D conversion.

Experiments follow a uniform format and are divided into three sections:

1. The experimental laboratory data. Data tables are marked with the letter E.
2. Required results. The experimental data is analyzed and converted to general logic levels. Corresponding tables are marked R.
3. Discussion. In this section the student is required to answer a series of questions which are based upon the experimental results and which integrate the data and digital theory. Corresponding tables are marked with the letter D.

Many types of IC logic families have been developed. Any one family could be used to provide a complete set of experiments. This manual provides experience with TTL and the Operational Amplifier.

Whenever possible, an experiment starts with the basic building of a logic operation. Once the principle has been developed, it is replaced by an MSI unit. This provides the student with an understanding of the logic and gives him a feeling for the trend in integrated circuits towards MSI and LSI and, in addition, simplifies the wiring of more complex circuits.

Discrete active components are used briefly in three experiments. Diodes are used in experiments 1, 11 and 12, and transistors in experiment 11 to develop some of the principles of the astable multivibrator.

All of the ICs used in this manual are in the 14 pin or 16 pin dual-in-line (DIP) construction. In appendix G are the constructional details of a laboratory IC socket and switch bank arrangement which readily lends itself to the wiring and breadboarding needs of a laboratory and to squad organization and participation. The IC socket is mounted on 1/8" aluminum channel and the pins are connected to universal 5-way binding posts. This permits the use of stackable banana plug leads for interconnection.

All of the experiments can be performed with 6 IC sockets and 2 switch banks. The cost of the material for the sockets and switch banks is approximately $100.00. At the present writing the cost of the ICs needed to perform all the experiments is less than $25.00. A complete set-up can be provided at a basic cost of less than $125.00.

I wish to express my appreciation to Richard Brown and John Gappa, laboratory technicians at the Staten Island Community College, for their suggestions and assistance, particularly with respect to the socket and switch bank construction, to the students at the SICC for their help and suggestions, to Dr. Irving Kosow, particularly with respect to the format, and to Mrs. Jean Johnson for her excellent typing and proofreading of the manuscript.

The ICs used in the experiments are all in the dual in-line 14 pin or 16 pin configuration. A convenient method of storage is in the plastic shipping container used by the IC manufacturers for shipping production quantities of ICs to the users.

The ICs used in the experiments (as is true for semiconductors in general) do not have a great over voltage tolerance before their dissipation is exceeded. The procedure suggested in appendix A, of calibrating the CRO and checking the power supply voltage against it, if carefully followed at the beginning of each experiment will protect the ICs against over voltage and at the same time provide some additional CRO experience.

Voltage measurements should be made to within 0.1 volt.

The ICs used have short internal connections and high F_T. If allowed to remain in the active region there is a great possibility of parasitic oscillation. This can occur in obtaining the transfer characteristics in experiment four. The by-pass capacitor at the gate input is a parasitic oscillation suppressor. For the same reason a pulse generator should have a short rise and fall time especially when toggle flip-flops are being used, to prevent the build up of oscillations and a resultant false count. Should there be any such suspicion, the Schmitt Trigger of experiment 12 can be used to shorten the rise and fall times. Schmitt Trigger IC types 7413 or 7414 can be used to reduce rise and fall times of pulse generators.

If the pulse generator is square wave centered around zero voltage it cannot be used with the ICs since they will not accept negative going gate input voltages. A small signal diode can be used across the pulse generator to clip the negative going portion of the wave.

While the wiring in some of the experiments is somewhat complex, a wiring layout as close as possible to the circuit drawing will help in trouble shooting.

Students should be encouraged to write the values of the logic variables at each point of the logic diagrams in the manual.

If a dual beam or dual trace CRO is available, the bistable action of experiment 8 can be best demonstrated if the beams are displaced horizontally from each other or one beam is operated with a lower brightness.

Flip-Flop Output Measurements — Precaution

A long lead from a bistable, or counter, or intermediate outputs of a counter, or monostable multivibrator may cause undesired loading and prevent the bistable from toggling or dividing properly, or the monostable from functioning properly. If such difficulties are encountered, a composition 2200Ω decoupling resistor at the IC end of the lead can eliminate the loading problems. The lead length from the 2200Ω resistor should be kept reasonably short. This decoupling may be needed both for the CRO vertical input and for external triggering.

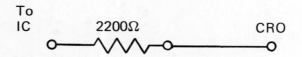

Integrated Circuit Types and Experiments in Which They Are Used

	741C	7400	7402	7403	7404	7405	7410	7420	7432	7441A	7472	7476	7481A/MC4005P	7486	7490
1. Basic Logic Functions		1	1	1			1	1							
2. Boolean Algebra		1		1	1			1							
3. DeMorgan's Theorem		1	1	1	1		1								
4. TTL NAND/NOR Gates		1		1											
5. Exclusive OR		1	1	1			1							1	
6. Full Adder/Full Subtractor		1	1	1		1	1							1	
7. Bistable		1	1								1				
8. Binary Counters		1					1				3				
9. Divide by N Counters											4				1
10. Shift Registers		1									1	3			
11. Pulse Forming		1		1											
12. Pulse Shaping/Schmitt Trigger				1											
13. Decoding/Encoding					1			2		1		1			1
14. Random Access Memories (RAM)		2		1									1		
15. Operational Amplifier	2														
16. D/A and A/D Conversion	2				1	1									1

IC PIN CONNECTIONS AND IDENTIFICATION

The ICs used in this manual are made in the dual in-line case. The IC pin terminals progress in a counterclockwise direction as seen from the top side away from the pins. The ICs are in either 14-pin TO-116 (EIA designation) or 16-pin configurations. In these ICs pin 1 is located by an identifying symbol, or the location of pins 1 and 14 (14 pin IC) or pins 1 and 16 (16 pin IC) are identified by an index notch at the end of the case where these pins are located. This is illustrated in the figure below.

Top View (away from pins)

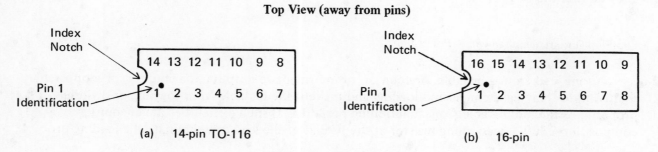

(a)   14-pin TO-116

(b)   16-pin

Dual in-line pin location.

OBJECT

(a)   To study the basic logic functions AND, OR, INVERT, NAND, NOR.

(b)   To study the representation of these functions by truth tables, logic diagrams, and Boolean algebra.

INTRODUCTORY THEORY

In electronic logic circuits inputs and outputs occur as voltage levels. These inputs and outputs are dual valued or dual leveled. To provide a common basis for comparison it has become usual to represent the two levels symbolically as 1 or 0. In one circuit the 1 might be +20 volts and the 0 might be −10 volts. In some other circuit the 1 might be +15 volts and the 0 might be +1 volt. Despite the difference in voltage levels, a basic operation will be the same using either pair of voltages.

AND: A multi-input circuit in which the output is a 1 only if all inputs are 1.

OR: A multi-input circuit in which the output is a 1 when any input is a 1.

INVERT: The output is 0 when the input is 1, and the output is 1 when the input is 0.

NAND: AND followed by INVERT.

NOR: OR followed by INVERT.

Truth table: Representation of the output logic levels of a logic circuit for every possible combination of levels of the inputs. This is best done by means of a systematic tabulation.

EQUIPMENT REQUIRED

CRO, dc coupled and calibrated.
dc power supply, +5 volts at 50 mA,
2 silicon small signal diodes, 1N457 or equivalent.
1   2.2 kΩ ± 10% composition resistor.
IC type 7400 quad 2–input NAND gate.
IC type 7402 quad 2 –input NOR gate.
IC type 7404 hex inverter.
IC type 7420 dual 4–input NAND gate.
IC type 7432 quad 2–input OR gate.
Switch bank, five switches per bank.

IC Manufacturers' Part Numbers.

Type	Motorola	Fairchild	Texas Instruments	National Semiconductor
7400	MC7400P MC7400L	7400PC 7400DC	SN7400N SN7400J	DM7400N
7402	MC7402P MC7402L	7402PC 7402DC	SN7402N SN7402J	DM7402N
7404	MC7404P MC7404L	7404PC 7404DC	SN7404N SN7404J	DM7404N
7420	MC7420P MC7420L	7420PC 7420DC	SN7420P SN7420J	DM7420N
7432		7432PC 7432DC	SN7432P SN7432P	DM7432N

IC PIN CONNECTIONS

Each of these ICs are in a 14-pin dual in-line case. The base pins progress in a counter-clockwise direction as seen from the side away from the pins, as shown in figure 1-1. Pin 1 is located by an identifying symbol, or the location of pins 1 and 14 are identified by an index notch at the end of the case where pins 1 and 14 are located.

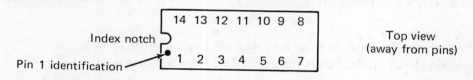

Index notch

Pin 1 identification

14 13 12 11 10 9 8

1 2 3 4 5 6 7

Top view
(away from pins)

**Fig. 1-1. IC pin location,
14-pin dual in-line (TO-116) case.**

PRELIMINARY PRECAUTIONS

The ICs used in this experiment are fragile and have a small voltage overload margin. It is extremely important that the exact required voltage be applied. The power supply voltage should be the last connection made. Before connections to the power supply are made, its voltage should be checked both against its own voltmeter and against the calibrated CRO.

All measurements in this experiment are made with a calibrated CRO. *Before making any measurements, calibrate the CRO* (see Appendix A). After the CRO is calibrated, use it to check the power supply voltage. If a disagreement occurs between the power supply voltmeter reading and the CRO measurement, call the instructor.

EXPERIMENTAL PROCEDURE

For each part of the experiment apply the indicated voltage and make voltage measurements at the points indicated to complete the tables. Use the CRO to measure the voltage. Use a sensitivity of 1 volt/div and make measurements to within 0.1 volt.

1. AND Function with Diodes

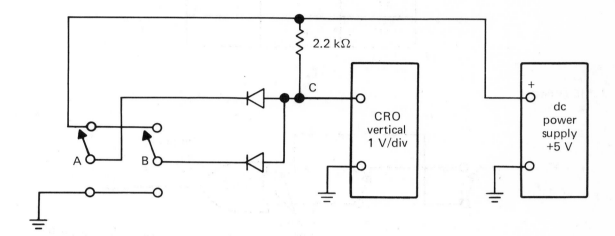

Fig. 1-2. AND gate.

Table 1-1E. Diode AND gate.

A	B	C
0	0	
0	+5	
+5	0	
+5	+5	

2. OR Function with Diodes

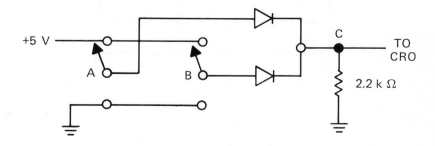

Fig. 1-3. OR gate.

3

Table 1-2E. Diode OR gate.

A	B	C
0	0	
0	+5	
+5	0	
+5	+5	

3. IC OR gate

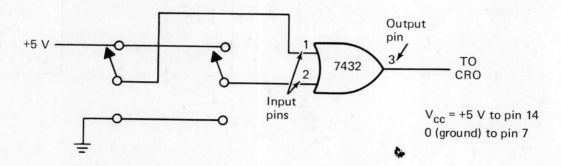

Fig. 1-4. IC OR gate.

V_{cc} = +5 V to pin 14
0 (ground) to pin 7

Note: The IC type 7432 has four 2-input OR gates within its package. Only one of these gates is used in this experiment. It is the gate whose input pins are pins 1 and 2 and whose output pin is pin 3. See Appendix D for complete logic diagram and pin connections.

Table 1-3E. IC OR gate.

Pin 1	Pin 2	Pin 3
0	0	
0	+5	
+5	0	
+5	+5	

4

4. INVERT

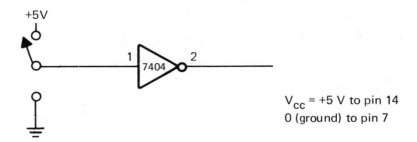

V_{cc} = +5 V to pin 14
0 (ground) to pin 7

Fig. 1-5. IC inverter.

Table 1-4E. INVERT.

Pin 1	Pin 2
0	
+5	

5. OR + INVERT = NOR

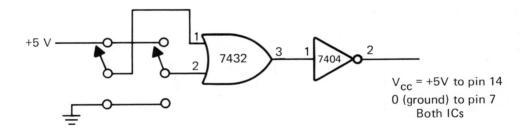

V_{cc} = +5V to pin 14
0 (ground) to pin 7
Both ICs

Fig. 1-6. OR + INVERT.

Table 1-5E.

7432		7404
Pin 1	Pin 2	Pin 2
0	0	
0	+5	
+5	0	
+5	+5	

6. NOR

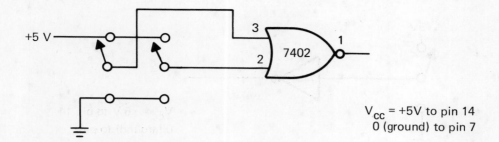

V_{cc} = +5V to pin 14
0 (ground) to pin 7

Fig. 1-7. NOR.

Table 1-6E.

Pin 3	Pin 2	Pin 1
0	0	
0	+5	
+5	0	
+5	+5	

7. 2-INPUT NAND

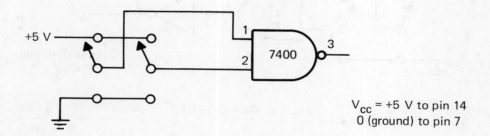

V_{cc} = +5 V to pin 14
0 (ground) to pin 7

Fig. 1-8. NAND.

Table 1-7E.

Pin 1	Pin 2	Pin 3
0	0	
0	+5	
+5	0	
+5	+5	

6

8. 4-Input NAND

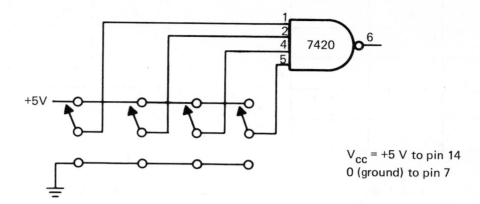

Fig. 1-9. 4-input NAND.

Table 1-8E.

Pin 1	Pin 2	Pin 4	Pin 5	Pin 6
0	0	0	0	
+5	0	0	0	
+5	+5	0	0	
+5	+5	+5	0	
+5	+5	+5	+5	

REQUIRED RESULTS

For each part of the experimental results complete the following truth tables. Use 1 and 0, defining logic 1 as a voltage greater than 2.5 volts and logic 0 as a voltage less than 1.0 volt. The table has been completed for part 1. Table 1-1R corresponds to the data of Table 1-1E, and so on. All tables should have 1s and 0s.

Table 1-1R.

A	B	C
0	0	0
0	1	0
1	0	0
1	1	1

Table 1-2R.

A	B	C

7

Table 1-3R.

Pin 1	Pin 2	Pin 3

Table 1-4R.

Pin 1	Pin 2

Table 1-5R.

7432		7404
Pin 1	Pin 2	Pin 2

Table 1-6R.

Pin 3	Pin 2	Pin 1

Table 1-7R.

Pin 1	Pin 2	Pin 3

Table 1-8R.

Pin 1	Pin 2	Pin 4	Pin 5	Pin 6

DISCUSSION

1. Figure 1-2 is titled an AND gate. Based on the experimental data of Tables 1-1E and 1-1R, explain why it is an AND gate.

2. Explain the values of the voltages measured in part 1 as tabulated in column C of Table 1-1E.

3. Figure 1-3 is titled an OR gate. Based on the experimental data of Tables 1-2E and 1-2R, explain why it is an OR gate.

4. Explain the values of the voltages measured in part 2 as tabulated in column C of Table 1-2E.

5. Based on the data of Table 1-3E and truth Table 1-3R, what kind of gate is IC type 7432?

6. Based on Table 1-4E and truth Table 1-4R, explain the meaning of "INVERT."

7.

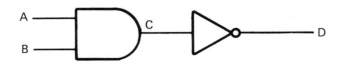

Fig. 1-10.

(a)　What type of logic function will occur between A, B, and C in figure 1-10?

(b)　What type of logic function will occur between A, B, and D in figure 1-10?

(c)　Write a truth table in the columns of Table 1-1D between A, B, C, and D, using 1s and 0s for figure 1-10.

Table 1-1D.

A	B	C	D
0	0		
0	1		
1	0		
1	1		

8.　Express the Boolean equations between inputs and outputs in Table 1-2D for each part of the experiment. Use letter inputs and outputs where they are used. With ICs the Boolean equation between input and output is expressed using the pin numbers. For example, suppose we had a 2-input AND gate with inputs at pins 4 and 5 and output at pin 6. The IC specification data sheet would be written as $4 \cdot 5 = 6$ (rather than pin 4 $\cdot$ pin 5 = pin 6)

Table 1-2D.

Part	Boolean Equation
1	
2	
3	
4	
5	
6	
7	
8	

10

BOOLEAN ALGEBRA AND SIMPLIFICATION OF LOGIC EQUATIONS

OBJECT

To study methods of representing and simplifying logic equations by Boolean algebra.

EQUIPMENT REQUIRED

CRO, dc coupled and calibrated.
dc power supply, +5 volts at 50 mA.
Square Wave Generator (SWG), 10 kHz, 0 to +5 volts.
IC type 7400 quad 2-input NAND gate.
IC type 7404 hex inverter.
IC type 7405 hex inverter, open collector.
IC type 7432 quad 2-input OR gate.
Switch bank, 5 switches per bank.
2 – 5.6 kΩ 10% composition resistors.

IC Manufacturers' Part Numbers.

Type	Motorola	Fairchild	Texas Instruments	National Semiconductor
7400	MC7400P MC7400L	7400PC 7400DC	SN7400N SN7400J	DM7400N
7404	MC7404P MC7404L	7404PC 7404DC	SN7404N SN7404J	DM7404N
7405	MC7405P MC7405L	7405PC 7405DC	SN7405N SN7405J	DM7405N
7432		7432PC 7432DC	SN7432N SN7432J	DM7432N

PRELIMINARY PROCEDURE

CRO

To obtain the correct wave shapes it is necessary to use **external triggering** from the SWG. If a dual-beam or dual-trace CRO is being used, one of the traces can be used to monitor the SWG output while viewing the wave shapes at the required points.

View the output of the 10-kHz SWG on the CRO and adjust its output to +5 volts. Use EXT TRIGGER from the SWG as shown in figure 2-1a and use NEG, AUTO TRIGGER. Adjust the CRO horizontal sweep rate so that each cycle occupies two horizontal divisions as shown in figure 2-1b.

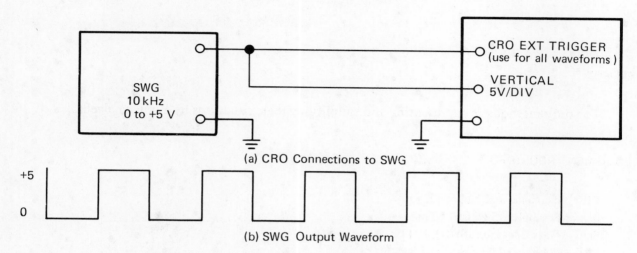

(a) CRO Connections to SWG

(b) SWG Output Waveform

Fig. 2-1. Square Wave Generator (SWG) output waveform.

Power Supply

Check the power supply voltage against the calibrated CRO. See Appendix A.

EXPERIMENTAL PROCEDURE

For all ICs in this experiment: V_{cc} = +5 Volts to pin 14, 0 (ground) to pin 7

There are two procedures in this experiment. In one, a voltage table must be completed and in the other, waveforms are viewed on the CRO and sketched on graph paper.

When a table must be completed, use the CRO to make voltage measurements at the points indicated to complete the table.

Where a signal input is applied from the SWG, view the waveforms on the CRO and sketch the waveforms at the indicated points on the graph paper provided with each part of the experiment. Use EXT. TRIG. to the CRO as shown in figure 2-1(a) always.

1(a)

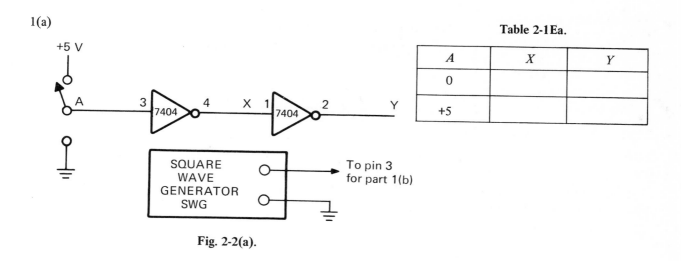

Fig. 2-2(a).

Table 2-1Ea.

A	X	Y
0		
+5		

1(b)   In figure 2-2(a) disconnect pin 3 from the switch and connect pin 3 to the SWG. View the waveforms at the points indicated and sketch on figure 2-2(b).

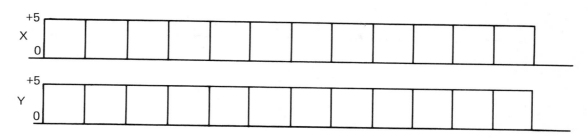

Fig. 2-2(b).

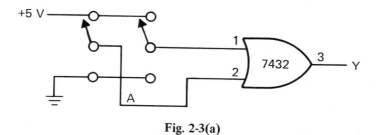

Fig. 2-3(a)

2(a)

Table 2-2Ea. Pin 1 connected to +5

A	Y
0	
+5	

2(b)

Table 2-2Eb. Pin 1 connected to ground

A	Y
0	
+5	

2(c)   In figure 2-3(a) disconnect pin 2 from the switch and connect pin 2 to the SWG. View and sketch the waveform on figure 2-3(b).

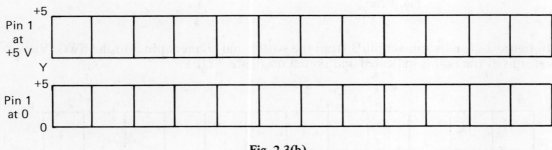

Pin 1
at
+5 V

Y

Pin 1
at 0

Fig. 2-3(b).

2(d)

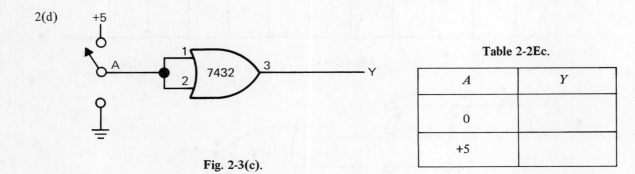

Fig. 2-3(c).

Table 2-2Ec.

A	Y
0	
+5	

2(e)   In figure 2-3(c) disconnect pins 1 and 2 from the switch and connect pins 1 and 2 to the SWG. Sketch the waveform on figure 2-3(d).

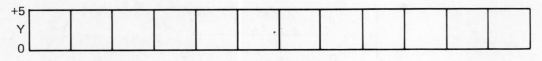

Fig. 2-3(d).

14

2(f)

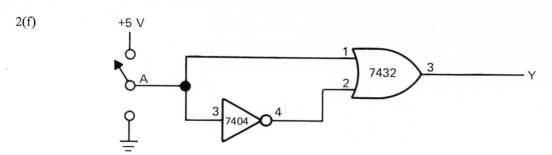

Fig. 2-3(e).

Table 2-2Ed.

A	Y
0	
+5	

2(g)   In figure 2-3(e) disconnect A from the switch and connect A to the SWG. View and sketch the waveforms on figure 2-3(f).

Note: A *glitch* (spurious short-duration pulse) may appear at Y. If it appears it should be shown on the waveform. However, its effect is to be ignored in the discussion section of the experiment.

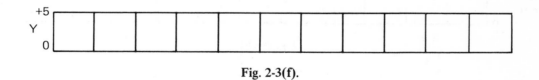

Fig. 2-3(f).

3(a)

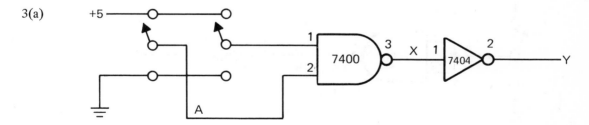

Fig. 2-4(a).

Table 2-3Ea. Pin 1 connected to +5.

A	X	Y
0		
+5		

3(b)

Table 2-3Eb. Pin 1 connected to ground.

A	X	Y
0		
+5		

3(c)   In figure 2-4(a) disconnect pin 2 (type 7400) from the switch and connect it to the SWG. View and sketch the waveforms at the points indicated on figure 2-4(b).

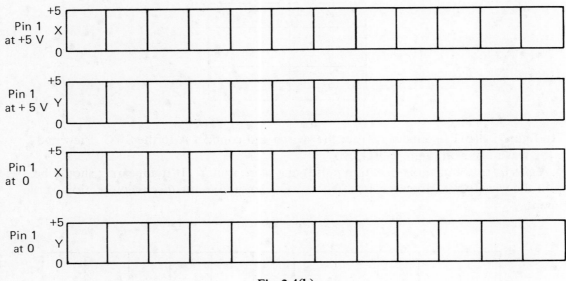

Fig. 2-4(b).

3(d)

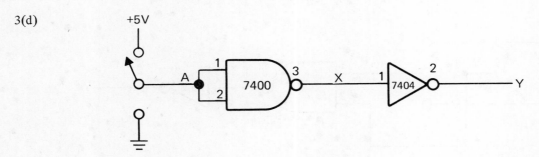

Fig. 2-4(c).

Table 2-3Ec.

A	X	Y
0		
+5		

16

3(e)   In figure 2-4(c) disconnect A from the switch and connect A to the SWG. View and sketch the waveforms at the points indicated on figure 2-4(d).

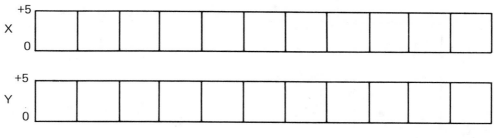

Fig. 2-4(d).

3(f)

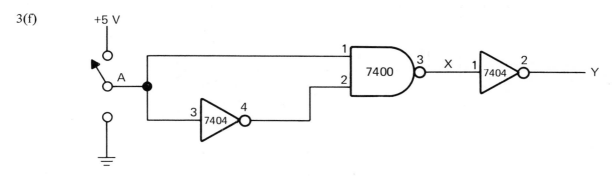

Fig. 2-4(e).

Table 2-3Ed.

A	X	Y
0		
+5		

3(g)   In figure 2-4(e) disconnect A from the switch and connect A to the SWG. View and sketch the waveforms at the points indicated on figure 2-4(f).

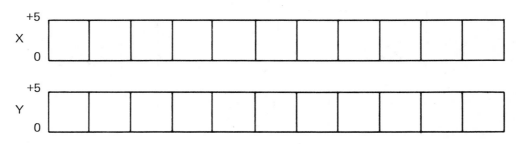

Fig. 2-4(f).

17

Make the voltage measurements at the required points to complete Tables 2-4Ea and 2-4Eb for the configurations of figures 2-5(a) and 2-5(b).

4(a)

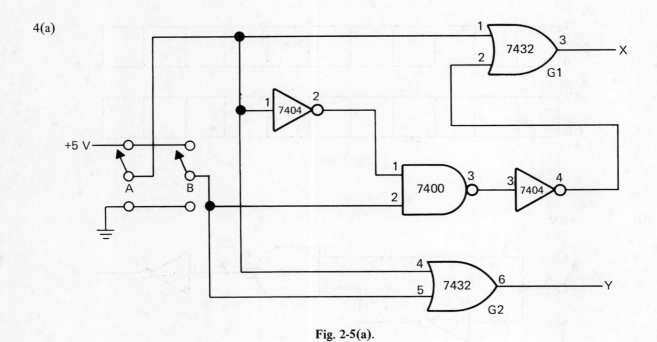

Fig. 2-5(a).

Table 2-4Ea.

A	B	$\bar{A}B$	X	Y
0	0			
0	+5			
+5	0			
+5	+5			

4(b)

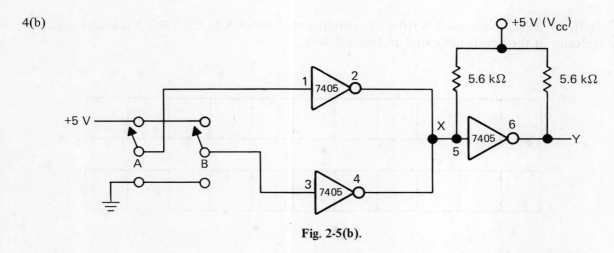

Fig. 2-5(b).

18

Table 2-4Eb.

A	B	X	Y
0	0		
0	+5		
+5	0		
+5	+5		

REQUIRED RESULTS

Complete the following tables corresponding to the voltage table of the experimental data. Assume positive logic in which logic $1 > 2.5$ volts and logic $0 < 0.5$ volt. All of the tables should have 1s and 0s.

Table 2-1Ra.

A	X	Y

Table 2-2Ra.

A	Y

Table 2-2Rb.

A	Y

Table 2-2Rc.

A	Y

Table 2-2Rd.

A	Y

Table 2-3Ra.

A	X	Y

Table 2-3Rb.

A	X	Y

Table 2-3Rc.

A	X	Y

Table 2-3Rd.

A	X	Y

Table 2-4Ra.

A	B	$\overline{A}B$	X	Y

Table 2-4Rb.

A	B	X	Y

19

1.(a)   What is the function of the inverter between X and Y in part 3?

2.(a)   For parts 1, 2 and 3, compare your data at Y with that at A. The Y data will be either 1, 0, or the same as A. Enter the results of Y as either 1, 0 or A in column Y. (Neglect any glitches which may have appeared in the waveforms). For example in parts 1(a) and 1(b) Y is the same as A and A has been entered in column Y and in parts 2(a) and 2(c), Y is always equal to 1. One (1) has been entered into column Y.

(b)   There is a Boolean equation which expresses the relationship between the input A and the output Y for each row of Table 2-5D which has been demonstrated experimentally. This equation can be determined from the logic diagram. From the logic diagram and the experimental data, express this equation in the Boolean equation column. For example, in parts 1(a) and 1(b) the experimental data has shown that Y = A. The logic diagram is that of a double inversion. The Boolean equation is $\bar{\bar{A}} = A$ and it has been entered in Table 2-5D, and for parts 2(a) and 2(c), the Boolean equation is A + 1 = 1. This has been entered into Table 2-5D.

Table 2-5D.

Part	Y A, 1, or 0	Boolean Equation Between Input A or SWG and Y
1(a) and 1(b)	A	$A = \bar{\bar{A}}$
2(a) and 2(c), pin 1 at +5	1	A + 1 = 1
2(b) and 2(c), pin 1 at 0		
2(d) and 2(e)		
2(f) and 2(g)		
3(a) and 3(c), pin 1 at +5		
3(b) and 3(c), pin 1 at 0		
3(d) and 3(e)		
3(f) and 3(g)		

3.   For part 4(a),

(a)   What are the two inputs to gate G1 in terms of A and B?

20

(b)　What is the output X in terms of A and B?

(c)　What is the output Y in terms of A and B?

(d)　Based on your experimental results, what can we say (in Boolean terms) about the relationship between X and Y? (Express in terms of A and B.)

4.　For part 4(a), draw two Karnaugh maps and discuss your results for the inputs to (a) G1 and (b) G2.

5.　When the output of 2 gates (in this case inverters) are wired together, (WIRED-COLLECTOR logic) we obtain a form of logic known as WIRED-OR, COLLECTOR-OR, PHANTOM-OR or IMPLIED-AND, DOT-AND. The most popular is WIRED-OR. This connection is permitted if the output collector loads are resistors (Passive Pull-up). TTL logic which has an active pull-up (the output collector load is another transistor) does not permit this because the output transistor may be damaged. However, some TTL ICs, such as the type 7405 INVERTER used in part 4b, have an open output collector. This allows the use of a resistor as an output collector load. Two or more output collectors can be wired together using a common load resistor for WIRED-COLLECTOR logic. The primary advantage of this type of logic, as can be seen from the WIRED-OR designation, is the simple generation of OR/NOR logic without the need for an additional OR or NOR gate.

From the truth table of part 4b, what type of gate is formed by the two inverters at point X? At point Y? Write the Boolean equations for points X and Y and the inputs A and B.

6.　It is desired to gate (turn on and off) the output of a SWG by two separate gating signals. When both gating signals are at logic 1 (positive logic), the SWG output is transmitted. When the SWG output is not being transmitted, the output is to be at logic 1. Draw a logic diagram showing how to accomplish this using a NAND gate. Explain how the circuit operates.

7. Repeat D9 but change the diagram so that the output is to be at logic 0 when the SWG is not being transmitted and use a NOR gate. (Use inverters if necessary.) Explain how the circuit operates.

Experiment 3

DE MORGAN'S THEOREM

OBJECT

To simplify and modify Boolean logic equations by means of De Morgan's theorem.

EQUIPMENT REQUIRED

CRO, dc and calibrated.
dc power supply, +5 volts at 50 mA.
Switch bank, five switches per bank.
IC type 7400 quad 2-input NAND gate
IC type 7402 quad 2-input NOR gate
IC type 7404 hex inverter
IC type 7405 hex inverter open collector
IC type 7420 dual 4-input NAND gate
5.6 kΩ 10% composition resistor.

IC Manufacturers' Part Numbers.

Type	Motorola	Fairchild	*Texas Instruments*	*National Semiconductor*
7400	MC7400P MC7400L	7400PC 7400DC	SN7400N SN7400J	DM7400N
7402	MC7402P MC7402L	7402PC 7402DC	SN7402N SN7402J	DM7402N
7404	MC7404P MC7404L	7404PC 7404DC	SN7404N SN7404J	DM7404N
7405	MC7405P MC7405L	7405PC 7405DC	SN7405N SN7405J	DM7405N
7420	MC7420P MC7420L	7420PC 7420DC	SN7420N SN7420J	DM7420N

For all the ICs in this experiment: V_{cc} = +5 Volts to pin 14, 0 (ground) to pin 7

For each part of the experiment wire the circuits shown and use the CRO to make voltage measurements at the points indicated to complete the associated table.

As you perform the experiment, for each point on the logic diagram express the Boolean equation in terms of the input variables and *write it on the diagram.* For example in part 1 you should write at point U, $\overline{A+B}$ and at point X you should write A+B.

(1)

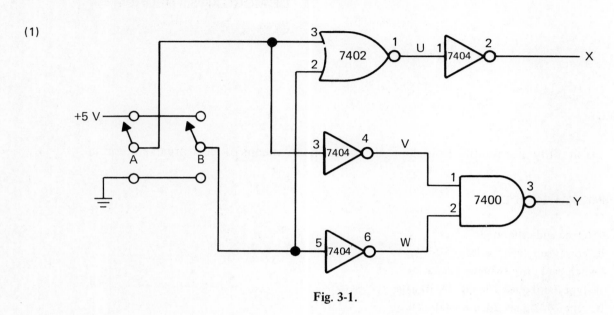

Fig. 3-1.

Table 3-1E.

A	B	U	V	W	X	Y
0	0					
0	+5					
+5	0					
+5	+5					

24

(2)

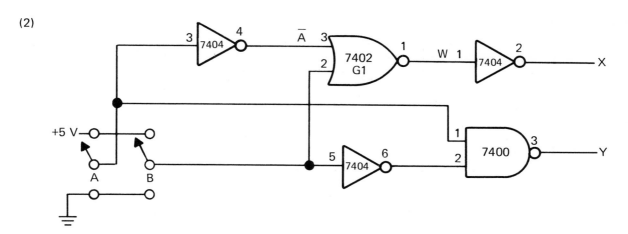

Fig. 3-2.

Table 3-2E.

A	$\bar{A}$	B	W	X	Y
0		0			
0		+5			
+5		0			
+5		+5			

(3)

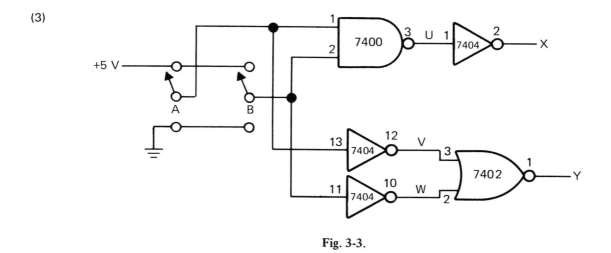

Fig. 3-3.

Table 3-3E.

A	B	U	V	W	X	Y
0	0					
0	+5					
+5	0					
+5	+5					

(4)

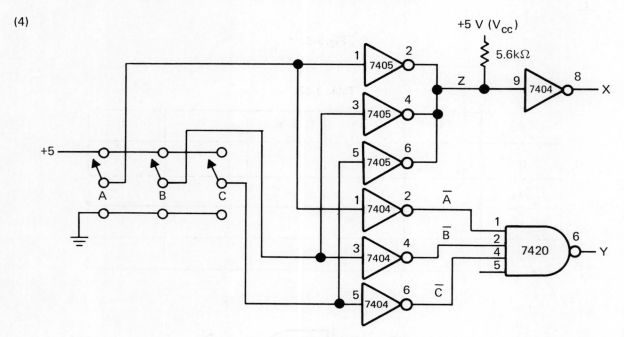

Fig. 3-4.

Table 3-4E.

A	B	C	$\overline{A}$	$\overline{B}$	$\overline{C}$	Z	X	Y
0	0	0						
0	0	+5						
0	+5	0						
0	+5	+5						
+5	0	0						
+5	0	+5						
+5	+5	0						
+5	+5	+5						

(5)

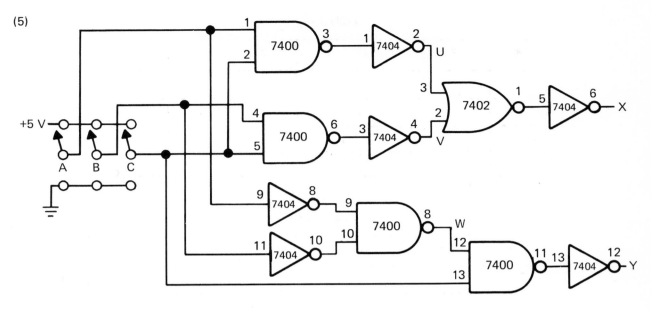

Fig. 3-5.

Table 3-5E.

A	B	C	U	V	W	X	Y
0	0	0					
0	0	+5					
0	+5	0					
0	+5	+5					
+5	0	0					
+5	0	+5					
+5	+5	0					
+5	+5	+5					

REQUIRED RESULTS

Assume positive logic, "1" ≡ > 2.5 volts, 0 ≡ < 0.5 volt.
Complete the following tables corresponding to each part of the experimental data obtained using logic "1" and "0."

Table 3-1R.

A	B	U	V	W	X	Y

Table 3-2R.

A	$\bar{A}$	B	W	X	Y

Table 3-3R.

A	B	U	V	W	X	Y

Table 3-4R.

A	$\bar{A}$	B	C	Z	X	Y

Table 3-5R.

A	B	C	U	V	W	X	Y

DISCUSSION

1. For part 1, in terms of the variable A and B, what are the Boolean algebra equations at U, V, W, X, and Y? Express the De Morgan relationship between A, B, X, and Y.

2. For part 2, in terms of the variables A and B and the inputs to G1, what is the De Morgan relationship between X and Y?

3. Draw Karnaugh maps for X and Y to explain the outputs X and Y of part 2.

4. For part 3, in terms of the variables A and B, what are the Boolean algebra equations at U, V, W, X, and Y? Express the De Morgan relationship between A, B, X and Y. Draw a Karnaugh map showing the relationship between A, B, X, and Y.

5. For part 4, in terms of the variables A, B, and C, what are the values at X, Y, and Z? Express the De Morgan relationship between A, B, C, X, Y, and Z. Explain the circuit operation at point Z.

6. For part 5, in terms of the variables A, B, and C, what are the values at U, V, W, X, and Y? Express the De Morgan relationship between A, B, C, X, and Y.

OBJECT

    a)   To study the operation of the TTL gate

    b)   To determine the loading rules for the TTL gate

    c)   To define the logic levels 1 and 0

    d)   To determine the nose immunity of the TTL gate

INTRODUCTORY THEORY

It is not essential to know the internal construction and component values within a gate (or IC). Digital ICs are frequently treated as black boxes and correctly perform logic function, providing their operating rules are understood and followed. However, for this experiment, where the object is to study the operation and rules of the 54/74 TTL (transistor-transistor-logic) series of ICs, the schematic diagrams of two types of 2-input NAND gates, the types 7400 and 7403 are given in figure 4-1.

Transistor Q_1 has two emitters at inputs A and B. These operate as a 2-input AND gate similar to the diode AND gate of experiment 1. Transistors Q_2 and Q_3 perform the operation of inversion. Transistor Q_2 acts as a driver for transistor Q_3 which is driven into saturation in the low level output state. In figure 4-1a, in the low level output state transistor Q_4 is cut off and in the high level output state, Q_4 conducts and acts as an emitter follower to provide a low-impedance output. This is called an active pull-up. Diode D is needed for biasing purposes. In the high level output state transistors Q_2 and Q_3 are cut off. In figure 4-1b, the collector of Q_3 is left open circuited. This is to allow a resistor to be connected from the collector of Q_3 to V_{CC} (passive pull-up) and permit wired-collector[a] logic.

Assume input B is at +5 volts and input A is held at a low level, for example, by another gate of the 54/74 series. Current flows through the 4K resistor and the base-emitter diode of Q_1 to input A. This current is sinked to ground through the output transistor of the grounding gate. We define this sinked current as a unit load (UL). A UL is sometimes expressed as an equivalent resistance referred to the supply voltage. However, there is a limit to the number of ULs that an output transistor can sink and this is determined by the gain of the output transistor and its associated circuitry. If we attempt to make the output transistor, such as Q_3, carry too heavy a current, it will come out of saturation, reduce ground level noise immunity [b] and may even make the gate ineffective if the output voltage exceeds the threshold voltage (see next paragraph) of

---

[a]Experiment 2, part 4 (b) and DISCUSSION − 5.

[b]See following theory on noise immunity − this experiment.

31

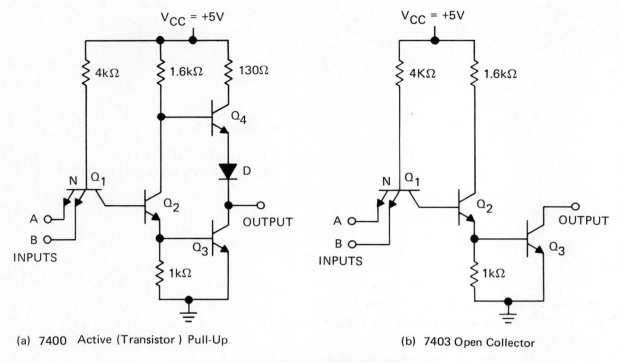

(a) 7400 Active (Transistor) Pull-Up   (b) 7403 Open Collector

Fig. 4-1. Type 7400 TTL 2-input NAND gates.

the next stages. The maximum number of ULs a gate can drive is called maximum fan-out and is specified on the IC manufacturer's data sheet. For the types 7400 and 7403, maximum fan-out is 10.

Assume in figure 4-1 that input B is at +5 volts and input A is at ground. With current flowing through the $4K\Omega$ input resistor, the voltage at point N at the base of transistor Q_1 is a diode drop (+0.7 volts) above ground. For current to flow into the base of Q_2, point N must be at +2.1 volts above ground (V_{BC} of Q_1 + V_{BE} of Q_2 + V_{BE} of Q_3). If the input voltage at point A is made more positive, point N will follow it but will be 0.7 volt more positive until N reaches 2.1 volts with A at 1.4 volts. At this input voltage level, the *threshold* voltage, the collector base diode of Q_1 conducts and the input emitter-base diode A cuts off with point N clamped at +2.1 volts. Transistors Q_2 and Q_3 and V_{CB} of Q_1 conduct only when inputs A and B are above the threshold voltage forming a 2-input NAND gate.

IC gates frequently have to drive capacitive loads. Consider an IC gate such as the type 7403 with resistor load driving a capacitive load such as shown in figure 4-2.

Figure 4-2a shows the output circuit. Figure 4-2b shows the input gate waveform. Figure 4-2c shows the output waveform (inverted from b) with the capacitor C = 0. Figure 4-2d shows the output waveform with a capacitor. When Q_3 is driven into saturation, it is a low resistance and discharges C so that the voltage across C $\approx$ 0. When Q_3 is turned off, the voltage at X can increase only as the capacitor C is charged through R_L. How fast it will charge is determined by the time constant $R_L C$.

For the gate to successfully drive another gate and cause it to switch, the voltage at X in figure 4-2a must rise above the threshold voltage for a following gate. If C is too large, or the switching rate is too fast, the voltage at X, as shown in figure 4-2d, cannot reach the threshold voltage before the next cycle arrives and the following gate will not have time to switch. To solve this problem, it is necessary that the time constant $R_L C$ be reduced. This can only be done by reducing R_L, but this additional current into Q_3 subtracts from the drive capability of the gate. TTL

32

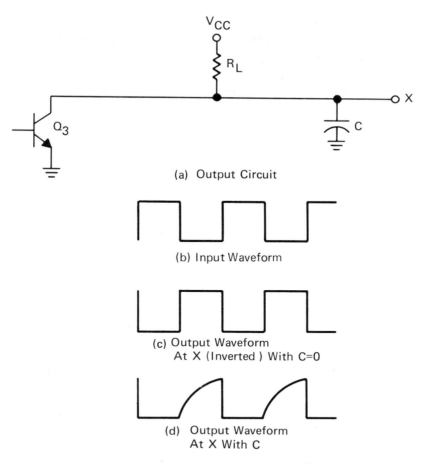

(a) Output Circuit

(b) Input Waveform

(c) Output Waveform
At X (Inverted) With C=0

(d) Output Waveform
At X With C

Fig. 4-2. IC gate with capacitive load.

gates such as the type 7400 with active pull-up solve this problem, but active pull-ups do not permit wired-collector logic.

Noise immunity, expressed in volts, is an important consideration in the design and choice of logic circuits. Noise immunity is a measure of the ability of a logic circuit to prevent unwanted noise signals from changing a desired logic level in response to the noise.

Consider the case of one gate driving a second gate of the same IC logic family as shown in figure 4-3.

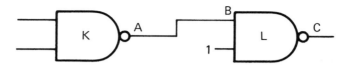

Fig. 4-3. IC NAND gate driving second gate.

Suppose that at A the output of gate K is high. This will make C, the output of gate L, low. Suppose noise in the form of

a) Electromagnetic coupling to the wire AB or

b) Electrostatic coupling to the wire AB

changes the voltage at B. How great a change in voltage downward can be tolerated before gate L is affected? The voltage at B can change to the point where the logic level of gate L changes. We have previously discussed the input voltage at which a gate changes from one level to the other and defined it as the threshold voltage. Similar factors must be considered when the output of gate K (point A) is at a low voltage and in this case at least an upward voltage change to the threshold voltage will cause the output of gate L to go through an undesired change. To evaluate these factors an IC input/output transfer characteristic is obtained which has the general shape shown in figure 4-4. This curve is applicable to both of the gates of figure 4-3.

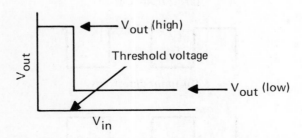

Fig. 4-4. Transfer characteristics of gates of Fig. 4-3.

From this curve we can determine the noise immunity which is expressed in the following equations.

$$\text{High Level Noise Immunity} = V_{out}\ (\text{high}) - V_{threshold} \qquad (4\text{-}1)$$

$$\text{Ground Level Noise Immunity} = V_{threshold} - V_{out}\ (\text{low}) \qquad (4\text{-}2)$$

Voltage changes in excess of these values will affect the output logic levels.

EQUIPMENT REQUIRED

CRO, dc coupled and calibrated. All voltage measurements are to be made with the CRO. Two-channel is advisable.

Square wave generator (SWG). Positive-going 5 volts at 10 kHz to 50 kHz. 500-Ω output impedance (max).

dc power supply, regulated to +5 volts at 100 mA.

dc milliammeter, 2-mA.

Capacitor, 0.033μF

Capacitor, 0.01μF

IC type 7400 quad 2-input NAND gate

IC type 7403 quad 2-input NAND gate, open collector

500Ω, 3 turn, 5 watt potentiometer

1000Ω composition resistor 10%

5.6 kΩ composition resistor 10%

Miscellaneous composition resistors from 100Ω to 10 kΩ

Type	Motorola	Fairchild	Texas Instruments	National Semiconductor
7400	MC7400P MC7400L	7400PC 7400DC	SN7400N SN7400J	DM7400N
7403	MC7403P MC7403L	7403PC 7403DC	SN7403N SN7403J	DM7403N

IC LOGIC DIAGRAMS

The logic diagrams, base pin connections, and loading rules for the 7400 IC and 7403 IC are shown in figure 4-5. Note that

(a) The input loading is 1 UL. This is shown in parentheses next to each input.

(b) The maximum fanout is 10 UL. This is shown in parentheses next to each output.

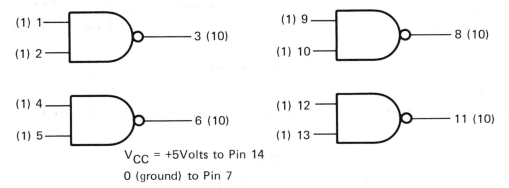

V_{CC} = +5Volts to Pin 14

0 (ground) to Pin 7

Fig. 4-5. Logic diagram, pin connections, and loading rules for types 7400 IC and 7403 IC.

EXPERIMENTAL PROCEDURE

Connect pin 14 to +5 volts and pin 7 to 0 (ground).

1. The Gate as an Inverter

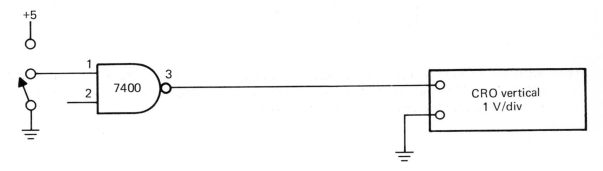

Fig. 4-6. The gate as an inverter.

Make the measurements at the terminal indicated to complete Table 4-1E.

Table 4-1E.

Voltage at Pin 1	Voltage at Pin 3
0	
+5	

2.   Gating Action

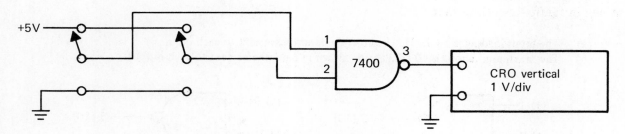

Fig. 4-7. 2-input gate.

(a)   Make the measurements at the terminals indicated to complete Table 4-2E.

Table 4-2E.

Voltage at Pin		Voltage at Pin
1	2	3
0	0	
+5	0	
0	+5	
+5	+5	

(b)   Make the measurements at the terminals indicated to complete Table 4-3E. An asterisk (*) in this table means that the terminal indicated is to be disconnected from the switch of part 2(a) and the terminal is to be left floating.

Table 4-3E.

Voltage at Pin		Voltage at Pin
1	2	3
+5	+5	
+5	*	
*	+5	
*	*	

* disconnected and floating

3. Loading Rules

(a) Determination of a UL. Measure the current in the circuit of figure 4-8.

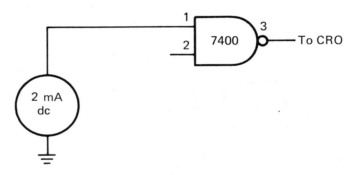

Fig. 4-8. UL determination.

$$I = \underline{\hspace{2cm}} \text{ mA}$$

(b) From the data of 3(a), calculate the resistance of a UL referred to +5 volts.

$$R_{1UL} = \underline{\hspace{2cm}} \Omega$$

(c) From the data of 3(b), calculate the equivalent current and resistance corresponding to 10 UL. This corresponds to the load a gate output would see if it were driving 10 gate inputs. DO NOT proceed until you have checked your answer with the instructor.

$$I_{10UL} = \underline{\hspace{2cm}} \text{ mA}$$

$$R_{10UL} = \underline{\hspace{2cm}} \Omega$$

(d) Increase the CRO vertical gain to 0.1 volt/div and carefully measure the voltage at pin 3 in the circuit of figure 4-9.

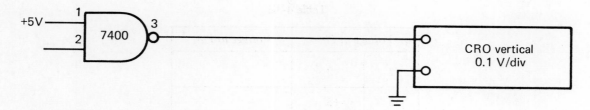

Fig. 4-9. Measurement of output voltage.

Voltage at pin 3 = _____ volts

(e)   Repeat the measurement of the voltage at pin 3 but with a resistor equal to 10 UL connected **between +5 volts and pin 3**. Use a composition resistor having the nearest 10 percent value to the value calculated in 3(c).

Voltage at pin 3 = _____ volts

4.   Transfer Characteristics

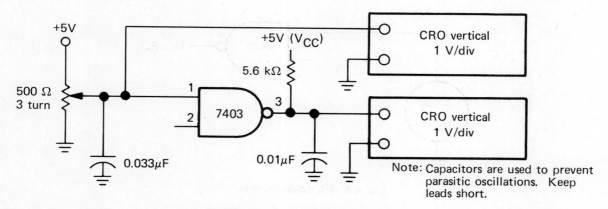

Fig. 4-10. Measurement of transfer characteristics.

(a)   The transfer characteristic is a curve showing the output voltage plotted against the input voltage. The transfer region is quite sharp and narrow, and careful measurements have to be made to obtain accurate results for a smooth curve. In making the measurements on the CRO, set the zero line on the CRO to allow a vertical deflection of at least 5 vertical divisions and use the VERT sensitivities shown in figure 4-10. Measure voltages to within 0.1 volt.

The following procedure is suggested in obtaining the data. Set the potentiometer in figure 4-10 to zero volts and measure the output voltage at pin 3. Now adjust the potentiometer to give an output voltage at pin 3 equal to approximately half the output voltage just obtained. Measure the input voltage at pin 1. Tabulate this set of input-output voltages in Table 4-4E and *plot* the point on figure 4-11. Vary the input potentiometer both above and below this mid-output voltage point and determine the maximum and minimum output voltages. Now vary the input potentiometer to obtain 3 approximately equally separated *output* voltage points between the mid-point and the maximum output voltage and 3 equally separated *output* voltage points between the mid-point and the minimum output voltage and measure the input voltage at each point. For each point *plot* it on figure 4-11 and tabulate the data in Table 4-4E. Complete the data by varying the input potentiometer so that the curve covers the **complete** input voltage range from 5 volts to zero. DO NOT PROCEED until your instructor has approved your data and plotted curve.

Table 4-4E.

$V_{pin\ 1}$	$V_{pin\ 3}$
Input Voltage	Input Voltage

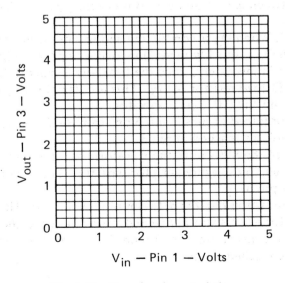

Fig. 4-11. Transfer characteristics.

39

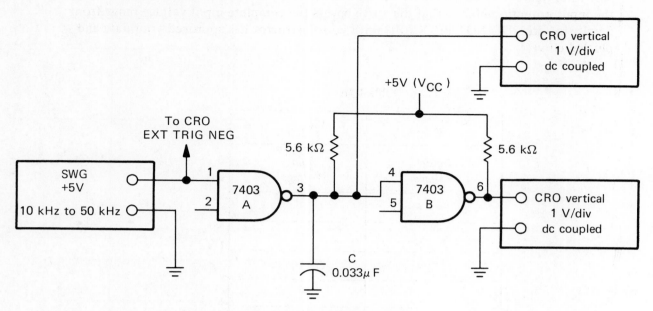

Fig. 4-12. Capacity loading.

In figure 4-12 the square wave generator (SWG) rapidly switches gates A and B, alternately turning the collector currents of Q_2 and Q_3 (see Fig. 4-1) on and off. Adjust the output of the SWG to provide a positive-going 5-volt peak signal at a frequency of 10 kHz.

(a) Disconnect the 0.033-μ F capacitor shown in figure 4-12 to make C = 0. Adjust the CRO sweep and triggering so that a steady pattern of approximately five cycles is obtained. Use *external* triggering for the CRO from the square wave generator as shown in figure 4-12. Sketch the waveform at IC output pins 3 and 6 on figure 4-13, for part 5(a).

(b) Connect the 0.033-μ F capacitor as shown in figure 4-12. This value of capacity is intended to simulate the effects that would occur if gate A and gate B were some distance apart and the connection between gate A and gate B was a long cable. The 0.033-μ F simulates the cable capacity.

Refer now to figure 4-1. When Q_3 is conducting and in saturation, pin 3 of gate A is at ground potential. When Q_3 is quickly made nonconducting by the SWG, the output at pin 3 cannot go to V_{CC} immediately, since the 0.033-μF capacitor has to charge up to V_{CC} through the 5.6-kΩ load resistor of Q_3. This is determined by the time constant of the 5.6-kΩ resistor and the 0.033-μF capacitor.

Sketch the waveforms obtained at pins 3 and 6 on figure 4-13 for part 5(b). The waveform should show the peak voltage reached at pin 3.

(c) Increase the frequency of the SWG until the waveform at pin 6 just disappears. Sketch the waveform at pins 3 and 6 on figure 4-13 for part 5(c). The waveform should show the peak voltage reached at pin 3.

(d) Connect a 1000-Ω composition resistor between pin 3 and V_{CC}. This is called a *pull-up* resistor. A pull-up resistor is a resistor connected from a point to the positive supply voltage. Sketch the waveforms at pins 3 and 6 on figure 4-13 for part 5(d). The waveform should show the voltage reached at pin 3.

40

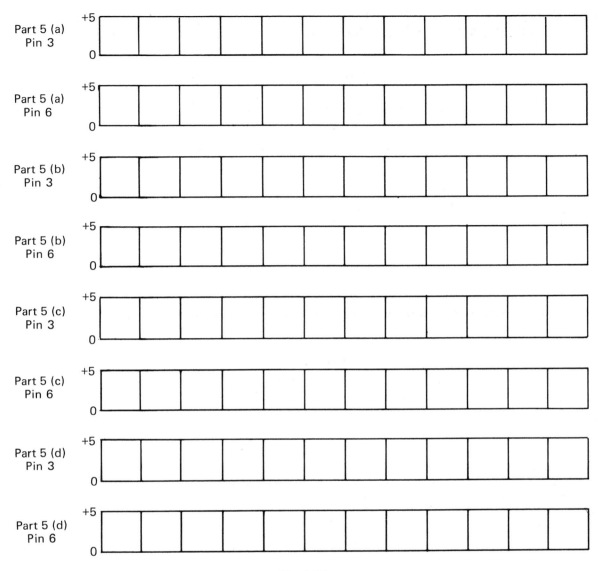

Part 5 (a) Pin 3

Part 5 (a) Pin 6

Part 5 (b) Pin 3

Part 5 (b) Pin 6

Part 5 (c) Pin 3

Part 5 (c) Pin 6

Part 5 (d) Pin 3

Part 5 (d) Pin 6

Fig. 4-13.

REQUIRED RESULTS

The numbers refer to the sections in the experimental procedure.

1. Using logic values $1 \equiv\, > 2.5$ volts and $0 \equiv\, < 0.5$ volt and based on the data of Table 4-1E, complete Table 4-1R. The table should have 1s and 0s.

Table 4-1R.

Pin 1	Pin 3

2. (a) Using logic values $1 \equiv > 2.5$ volts and $0 \equiv < 0.5$ volt for positive logic and based on the data of Table 4-2E, complete Table 4-2Ra. The table should have 1s and 0s.

Table 4-2Ra.

Pin 1	Pin 2	Pin 3

(b) Using logic values $1 \equiv < 0.5$ volt and $0 \equiv > 2.5$ volts for *negative* logic and based on the data of Table 4-2E, complete Table 4-2Rb. The table should have 1s and 0s.

Table 4-2Rb.

Pin 1	Pin 2	Pin 3

(c) Using logic values $1 \equiv > 2.5$ volts and $0 \equiv < 0.5$ volt and based on the data of Table 4-3E, complete Table 4-3R. The table should have 1s and 0s.

Table 4-3R.

Pin 1	Pin 2	Pin 3
	*	
*		
*	*	

DISCUSSION

1. Based on the data of Table 4-1R, draw the logic symbol for the gate operation between pins 1 and 3. (Omit power supply pin connections.)

2. (a) Explain, using the data of Tables 4-2E and 4-2Ra (positive logic) what type of gate this is. Draw the logic symbol representing its operation. Write the Boolean algebra equation for the circuit using the pin numbers as the Boolean variables.

   (b) Explain, using the data of Tables 4-2E and 4-2Rb (negative logic), what type of gate this is. Draw the logic symbol representing its operation. Write the Boolean algebra equation for the circuit using the pin numbers as the Boolean variables.

3. Based on the data of Tables 4-3E and 4-3R, what logic value, 1 or 0, does an unconnected input assume (positive logic)?

4. From a study of the IC circuit diagram of figure 4-1, compute the current that would flow through the emitter of input A, if A is connected to ground. Compare with your experimental data obtained in part 3(a).

5. Using transistor theory, why is the voltage measured in part 3(e) different from that in part 3(d)?

6. Referring to the graphical data of part 4 and figure 4-11, at what voltage applied to pin 1 does the voltage at pin 3 change most rapidly? Explain the reason for this value based on a study of figure 4-1.

7. The questions in this section are based on noise immunity and the data of part 4. Study the theory on noise immunity, carefully, before answering the questions. Use data of figure 4-11.

   (a) What gate input voltage applied to pin 1 corresponds to the middle of the sharp transition region of the gate output voltage of pin 3? (see Fig. 4-11). Call this the threshold voltage.

   (b) What is the high level noise immunity? Show calculations.

   (c) What is the ground level noise immunity? Show calculations.

   (d) Assume a minimum allowable ground-level noise immunity of 0.5 volt. Based on the threshold voltage of figure 4-11, what is the maximum possible voltage for the 0 voltage level (positive logic)? Show calculations.

   (e) Assume a minimum allowable high-level noise immunity of 0.5 volt. Based on the threshold voltage of figure 4-11, what is the minimum possible voltage for the 1 voltage level (positive logic)? Show calculations.

8. These questions refer to the data of part 5.
   (a) Explain the waveforms at pins 3 and 6 for part 5(a).

(b) Explain the waveforms at pins 3 and 6 for part 5(b). (Hint: Consider the effect of the 0.033-μF capacitor.)

(c) Carefully examine the amplitude of the wave shape at pin 3 for part 5(c) and compare it with the transfer curve and IC threshold voltage of part 4 and figure 4-11 of the experiment. From this, explain why the pattern at pin 6 disappeared.

(d) Explain the wave shapes of part 5(d). Why did the pattern reappear?

(e) What is the function of a pull-up resistor?

(f) How many ULs does the 1000-Ω pull-up resistor represent? With this pull-up resistor in the circuit, how many additional ULs can this IC drive?

(g) What are the advantages and disadvantages of a pull-up resistor?

9. Figure 4-12 is the logic diagram of a 2-input NAND gate followed by an inverter. Write the Boolean equation for figure 4-12 for positive logic. Write the Boolean equation for figure 4-12 for negative logic. This is to describe circuit operation between input pins 1, 2, and 6. Discuss the reasons for your answers.

OBJECT

(a)   To study methods of generating the EXCLUSIVE OR function.

(b)   The HALF-ADDER and HALF-SUBTRACTOR.

(c)   Binary comparators.

(d)   Parity generators.

INTRODUCTORY THEORY

Binary comparator: A circuit which compares two binary words bit by bit.

Parity generator: A circuit which determines whether there is an odd or even number of ls in a binary word.

EQUIPMENT REQUIRED

CRO, dc and calibrated.

dc power supply, +5 volts at 50 mA.

IC type 7400 quad 2-input NAND gate.

IC type 7402 quad 2-input NOR gate.

IC type 7404 hex inverter.

IC type 7420 dual 4-input NAND gate.

IC type 7486 quad EXCLUSIVE-OR gate.

2 switch banks, 5 switches per bank.

IC manufacturers' part numbers.

Type	Motorola	Fairchild	Texas Instruments	National Semiconductor
7400	MC7400P MC8400L	7400PC 7400DC	SN7400N SN7400J	DM7400N
7402	MC7402P MC7402L	7402PC 7402DC	SN7402N SN7402J	DM7402N
7404	MC7404P MC7404L	7404PC 7404DC	SN7404N SN7404J	DM7404N
7420	MC7420P MC7420L	7420PC 7420DC	SN7420N SN7420J	DM7420N
7486		7486PC 7486DC	SN7486N SN7486J	DM7486N

47

For all ICs in this experiment: V_{cc} = +5 Volts to pin 14, 0 (ground) to pin 7.

General: For each part of the experiment, wire the circuit shown. Make voltage measurements at the points designated to complete the tables.

1. EXCLUSIVE-OR generation

(a)

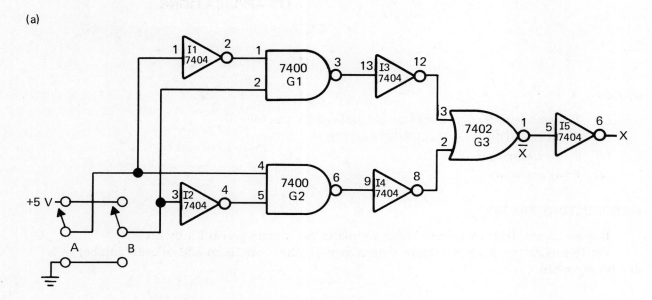

Fig. 5-1a. EXCLUSIVE-OR generator.

Table 5-1Ea.

A	B	$\overline{X}$	X
0	0		
0	+5		
+5	0		
+5	+5		

(b)

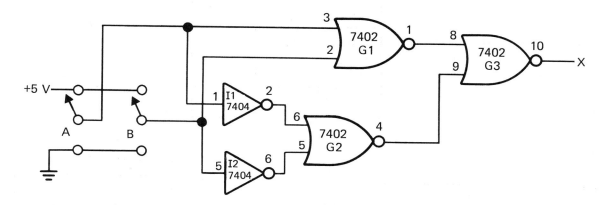

Fig. 5-1b. EXCLUSIVE-OR generator.

Table 5-1Eb.

A	B	X
0	0	
0	+5	
+5	0	
+5	+5	

(c)

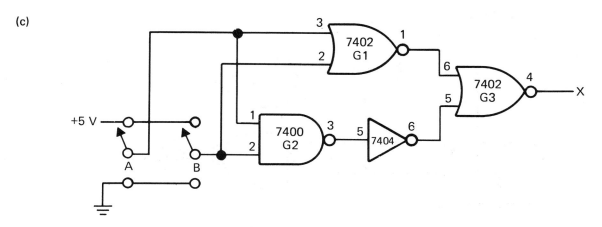

Fig. 5-1c. EXCLUSIVE-OR generator.

Table 5-1Ec.

A	B	X
0	0	
0	+5	
+5	0	
+5	+5	

(d)

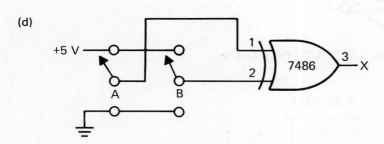

Fig. 5-1d. EXCLUSIVE-OR gate.

Table 5-1Ed.

A	B	X
0	0	
0	+5	
+5	0	
+5	+5	

2. HALF-ADDER, HALF-SUBTRACTOR

Since the HALF-ADDER and HALF-SUBTRACTOR require the EXCLUSIVE-OR as part of their logic, any configuration of part 1 could be used. The simplest is used in part 2.

(a)   HALF-ADDER, X + Y

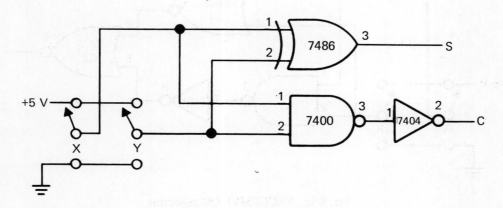

Fig. 5-2a. HALF-ADDER.

50

Table 5-2Ea.

X	Y	S	C
0	0		
0	+5		
+5	0		
+5	+5		

(b)   HALF-SUBTRACTOR, X − Y

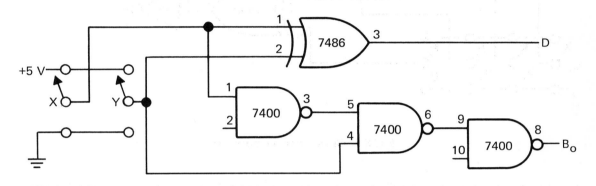

Fig. 5-2b. HALF-SUBTRACTOR.

Table 5-2Eb.

X	Y	D	B_O
0	0		
0	+5		
+5	0		
+5	+5		

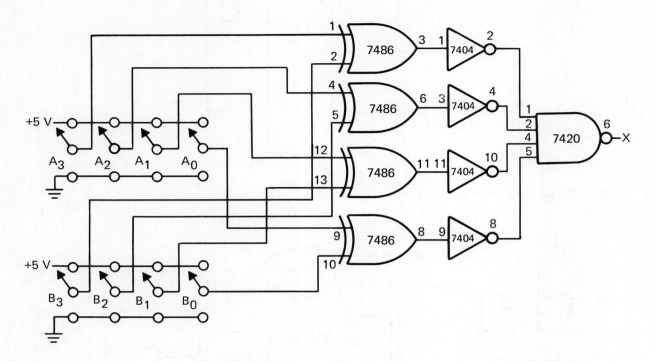

Fig. 5-3. Binary word comparator.

Set A_3, A_2, A_1, and A_0—the first binary word— to the following values and experimentally determine the values of B_3, B_2, B_1, and B_0 to make $X < 0.5$ volt.

Table 5-3E.

A_3	A_2	A_1	A_0	B_3	B_2	B_1	B_0
0	0	0	0				
0	+5	+5	0				
+5	+5	+5	0				
0	0	0	+5				
+5	0	+5	0				

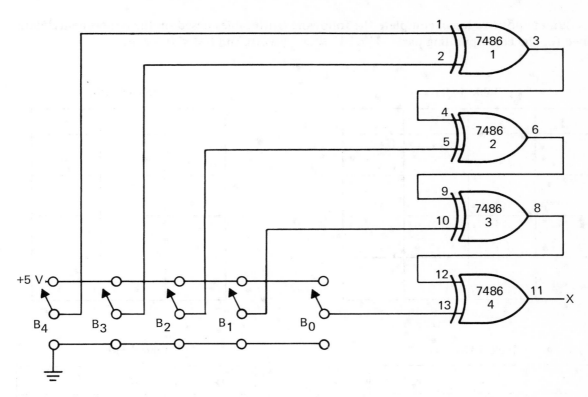

Fig. 5-4.

Set the binary words, B_4, B_3, B_2, B_1, B_0, to the values given in Table 5-4E and record the voltage at X.

Table 5-4E.

B_4	B_3	B_2	B_1	B_0	X
0	0	0	0	0	
0	+5	0	+5	0	
+5	0	+5	0	+5	
0	+5	+5	+5	0	
+5	+5	0	+5	+5	
+5	0	+5	+5	+5	
0	+5	0	0	0	
0	0	0	0	+5	

Assuming positive logic, complete the following truth tables based on the experimental data obtained for the corresponding parts. Use "1" $\equiv\, > 2.5$ volts and $0 \equiv\, < 0.5$ volt.

Table 5-1Ra.

A	B	$\overline{X}$	X

Table 5-1Rb.

A	B	X

Table 5-1Rc.

A	B	X

Table 5-1Rd.

A	B	X

Table 5-2Ra.

A	B	S	C

Table 5-2Rb.

X	Y	D	B_O

54

Table 5-3R.

A_3	A_2	A_1	A_0	B_3	B_2	B_1	B_0

(4)   Complete Table 5-4R for the experimental data. In the column marked T, total the number of 1s in the binary word B_4, B_3, B_2, B_1, B_0. In the column marked P indicate the parity of the word "ODD" or "EVEN."

Table 5-4R.

B_4	B_3	B_2	B_1	B_0	T Total 1s	P ODD or EVEN	X

DISCUSSION

1.   How does the circuit in part 1(a) generate the EXCLUSIVE-OR function? Discuss particularly with reference to the inputs to G3.

2. How does the circuit of part 1(b) generate the EXCLUSIVE-OR function? (Hint: Use De Morgan's theorem for the outputs of G_1 and G_2, and draw a Karnaugh map.)

3. How does the circuit of part 1(c) generate the EXCLUSIVE-OR function? (Hint: Determine the inputs to G_3 and draw a Karnaugh map.)

4. Write out a truth table for a half-adder and compare with the results of Tables 5-2Ea and 5-2Ra. Discuss.

5. Write out a truth table for a half-subtractor and compare with the results of Tables 5-2Eb and 5-2Rb. Discuss.

6. Explain the circuit operation of the BINARY word comparator of part 3. Refer to your experimental data. If necessary write a truth table.

7. Discuss and explain the circuit operation of the parity generator of part 4. Refer to your experimental data comparing columns T, P, and X. (Hint: Establish a pattern for parity generator outputs in the following manner. Assume you have a 2-bit parity generator. Determine its output. Now, do the same for a 3-bit system and then a 4-bit and then generalize.)

8. Why do we need word comparators and parity generators? Which parity is preferable, odd or even?

OBJECT

To study methods of generating circuits which perform the arithmetic operations of full addition and full subtraction.

EQUIPMENT REQUIRED

CRO, dc and calibrated.
dc power supply, +5 volts at 50 mA.
Switch bank, five switches.
IC type 7400 quad 2-input NAND gate.
IC type 7402 quad 2-input NOR gate.

IC type 7404 hex inverter.
IC type 7410 triple 3-input NAND gate.
IC type 7420 dual 4-input NAND gate.
IC type 7486 quad EXCLUSIVE-OR gate.

IC manufacturers' part numbers.

Type	Motorola	Fairchild	Texas Instruments	National Semiconductor
7400	MC7400P MC7400L	7400PC 7400DC	SN7400N SN7400J	DM7400N
7402	MC7402P MC7402L	7402PC 7402DC	SN7402N SN7402J	DM7402N
7404	MC7404P MC7404L	7404PC 7404DC	SN7404N SN7404J	DM7404N
7410	MC7410P MC7410L	7410PC 7410DC	SN7410N SN7410J	DM7410N
7420	MC7420P MC7420L	7420PC 7420DC	SN7420N SN7420J	DM7420N
7486		7486PC 7486DC	SN7486N SN7486J	DM7486N

For all the ICs in this experiment: V_{CC} = +5 Volts to pin 14, 0 (ground) to pin 7.

For each part, wire the circuit shown and make voltage measurements at the points indicated to complete the table.

1.   SUM in FULL ADDER and DIFFERENCE in FULL SUBTRACTOR

S for $X + Y + C_i$
D for $X - Y - B_i$

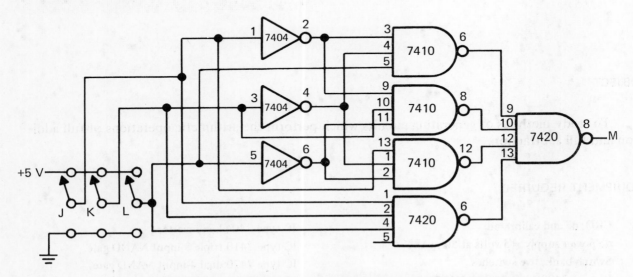

Fig. 6-1. Sum in **FULL ADDER** and **DIFFERENCE** in **FULL SUBTRACTOR**.

Table 6-1E.

Variable	J	K	L	M
Full Adder	X	Y	C_i	S
Full Subtractor	X	Y	B_i	D
	0	0	0	
	0	0	+5	
	0	+5	0	
	0	+5	+5	
	+5	0	0	
	+5	0	+5	
	+5	+5	0	
	+5	+5	+5	

2. C_O for FULL ADDER

C_O for $X + Y + C_i$

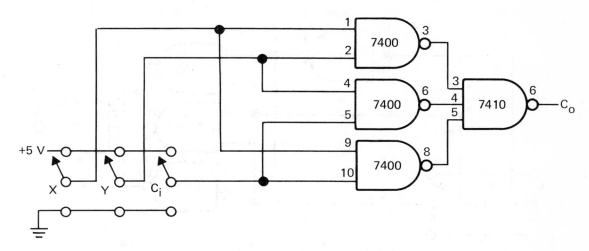

Fig. 6-2. C_O for FULL ADDER.

Table 6-2E.

X	Y	C_i	C_O
0	0	0	
0	0	+5	
0	+5	0	
0	+5	+5	
+5	0	0	
+5	0	+5	
+5	+5	0	
+5	+5	+5	

3. B_O for FULL SUBTRACTOR

B_O for $X-Y-B_i$

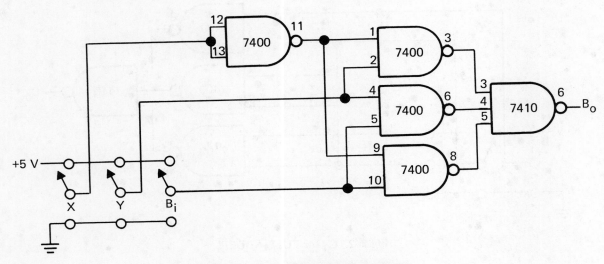

Fig. 6-3. B_O for FULL SUBTRACTOR.

Table 6-3E.

X	Y	B_i	B_O
0	0	0	
0	0	+5	
0	+5	0	
0	+5	+5	
+5	0	0	
+5	0	+5	
+5	+5	0	
+5	+5	+5	

62

4. FULL ADDER with HALF-ADDERS

(a) $X + Y + C_i$

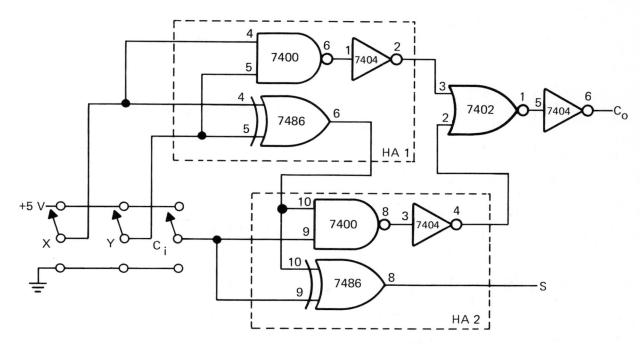

Fig. 6-4. FULL ADDER with HALF-ADDERS.

Table 6-4E.

X	Y	C_i	S	C_o
0	0	0		
0	0	+5		
0	+5	0		
0	+5	+5		
+5	0	0		
+5	0	+5		
+5	+5	0		
+5	+5	+5		

$$\begin{array}{r} X_1X_0 \\ +Y_1Y_0 \\ \hline SUM \end{array} \equiv \begin{array}{rr} X_1 & X_0 \\ +Y_1 & Y_0 \\ \hline C_{o1}S_1 & S_0 \end{array}$$

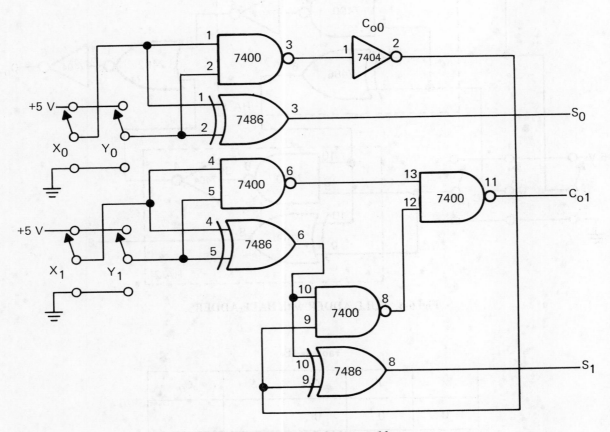

Fig. 6-5. 2-bit parallel binary adder.

(a)   The circuit of figure 6-5 is that of a 2-bit parallel binary adder, which performs the addition $X_1X_0 + Y_1Y_0 = C_{o1}S_1S_0$ when the numbers X_1X_0, Y_1Y_0, and the sum $C_{o1}S_1S_0$ are expressed in binary notation. Table 6-5Ea, has two sections, an Experimental Data section for laboratory data and a Binary Equivalents section for checking the experimental data against actual binary numeric addition. *Each row* of Table 6-5Ea *must be completed* before proceeding to the next row.   Make voltage measurements at the indicated points of figure 6-5 to complete the Experimental Data section.

The Binary Equivalents section of Table 6-5Ea must now be completed.   Using positive logic, $1 \equiv \, > 2.5$ volts, and $0 \equiv \, < 0.5$ volt, express the numbers X_1X_0 and Y_1Y_0 in binary notation.   Now, actually add them numerically and enter the sum as S_C (Sum Computed) in the S_C column.   Now express the sum $S_E = C_{o1}S_1S_0$ (Sum Experimental) from the Experimental Data section in the S_E column in binary notation. S_C and S_E should agree.   If they do, proceed to the next row.   If they do not agree, CALL THE INSTRUCTOR.

Table 6-5Ea.

Experimental Data								Binary Equivalents (in binary notation)			
X_1	X_0	Y_1	Y_0	C_{o1}	S_1	S_0		X_1X_0	Y_1Y_0	X_1X_0 $+Y_1Y_0$	$C_{o1}S_1S_0$ (experimental)
										S_C	S_E
0	0	0	0								
+5	+5	+5	+5								
+5	0	+5	+5								
0	+5	+5	0								

(b) Complete the experimental data in Table 6-5Eb.

Table 6-5Eb.

	X_1	X_0	Y_1	Y_0	C_{o1}	S_1	S_0
a	0	0	0	+5			
b	0	+5	0	+5			
c	0	+5	+5	+5			
d	+5	0	+5	0			
e	+5	0	+5	+5			

65

Using positive logic, complete the following tables corresponding to the experimental data. $1 \equiv > 2.5$ volts, $0 \equiv < 0.5$ volt.

Table 6-1R and 6-2R. FULL ADDER.

X	Y	C_i	S	C_O

Table 6-1R and 6-3R. FULL SUBTRACTOR.

X	Y	B_i	D	B_O

Table 6-4R.

X	Y	C_i	S	C_O
0	0	0		
0	0	+5		
0	+5	0		
0	+5	+5		
+5	0	0		
+5	0	+5		
+5	+5	0		
+5	+5	+5		

66

Complete the following table based upon the experimental data of part 5(b).

Table 6-5Rb.

Binary Notation				Decimal Equivalent		
X_1X_0	Y_1Y_0	X_1X_0 $+Y_1Y_0$	$C_{o1}S_1S_0$	X_1X_0	Y_1Y_0	X_1X_0 $+Y_1Y_0$
		S_C Sum Computed	S_E Sum Experimental			

DISCUSSION

1. Compare the full adder data of parts 1 and 2 and part 4 with a truth table for a full adder. Discuss your data. Explain the principle of obtaining full adder output for:
   (a) the circuitry of parts 1 and 2
   (b) the circuitry of part 4.

2. Does C_0 in part 2 for the full adder give correct results? Explain the equivalence of this circuit to the Boolean expression for

$$C_0 = XY\overline{C_i} + \overline{X}YC_i + XYC_i + X\overline{Y}C_i$$

(Hint: Use the Karnaugh Map.)

3. Compare the data of part 3 with the truth table for a full subtractor. Explain the equivalence of this circuit to the Boolean expression for

$$B_O = \overline{X}YB_i + \overline{X}\overline{Y}B_i + XYB_i + \overline{X}Y\overline{B_i}$$

(Hint: Use the Karnaugh Map.)

4. Draw a block diagram and a logic and wiring diagram for the full subtractor which uses two half-subtractors. Use the ICs necessary to perform this function and show the *pin numbers* on the logic diagram.

5. In the required results for part 5(b), does $S_C = S_E$?

6. Explain the operation of the 2-bit parallel binary adder of part 5.

7. Draw a block diagram showing a 3-bit parallel binary adder.

OBJECT

To study the characteristics and operation of various types of bistables.

INTRODUCTORY THEORY

THE BISTABLE: TYPES AND DEFINITIONS

The flip-flop (FF), bistable, or binary is a circuit with two (and *only* two) *stable states.* It remains in one of the stable states until a signal changes it into the second stable state and remains in the second stable state even upon removal of the signal. In a similar manner, a second signal can change the circuit from its second stable state back to its first stable state.

The standard FF symbol is shown in figure 7-1. It is common practice to designate the outputs as Q and $\overline{Q}$ or levels 1 and 0. A typical truth table is shown in Table 7-1. The FF is defined *only* by the states A and B given in Table 7-1.

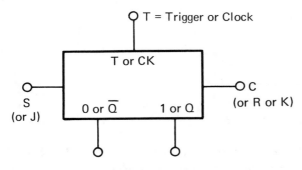

Fig. 7-1. The flip-flop (FF).

Table 7-1. FF States.

State	Q	$\overline{Q}$
A	0	1
B	1	0

In a transistorized bistable two transistors are *dc coupled* so that one transistor conducts while the second transistor is reverse biased and cannot conduct. The external circuit conditions and components of the nonconducting side are arranged to keep one transistor in conduction (most frequently in saturation), and the external circuit conditions and components of the conducting side in turn are arranged to keep the second transistor cut off.

71

Since the circuits for each transistor are identical, this symmetrical arrangement keeps either transistor in the *on* or *off* state, thereby providing a bistable arrangement. An external signal is required to change the FF from one state to another. One way to do this is to apply a signal to the *on* transistor in such a way as to make it nonconducting. This process, because of the dc coupling, simultaneously tends to bring the nonconducting side into conduction. A temporary unstable condition is quickly reached when both transistors are simultaneously conducting. One is trying to cut off and the second to conduct. The condition is regenerative, with the result that the transition, once initiated beyond a critical point, occurs very rapidly, and a stable reversal of state is reached.

The IC flip-flop is a modification of the basic transistorized FF, and IC gates take the place of the transistors previously discussed. Many different types of FFs have been designed. The basic types used with ICs will be discussed.

R-S or set-reset FF is defined as a bistable with SET and RESET inputs having the restriction that both inputs cannot be simultaneously energized because the resultant state of the bistable is indeterminate. The term CLEAR is frequently used in place of RESET. When made with two NAND or NOR gates, this FF is frequently called a *latch*. Figure 7-2 gives the logic symbol for the R-S flip-flop.

Fig. 7-2. Logic symbol for R-S flip-flop.

Gated R-S flip-flop is identical to an R-S flip-flop with respect to logic and truth table. Information can be transferred into the FF only when the gate is enabled. The terms CLOCKED, STROBE, and SYNCHRONOUS are frequently used in place of GATED. Figure 7-1 also shows the logic symbol of a gated R-S flip-flop.

D or data FF is a binary which has a single data (D) input and a clock input. Figure 7-3 gives the logic symbol for a D flip-flop.

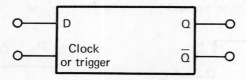

Figure. 7-3. D FF.

Master-slave FF consists of two dc-coupled, gated FFs. Information is transferred into the master section during one state of the clock pulse and from the master to the slave during the opposite state of the clock pulse. This arrangement isolates input and output and minimizes race problems.

A block diagram of an R-S master-slave FF is shown in figure 7-4. The notation means that information is entered into the master when the clock is 1, but the circle at the input to the slave means that information from the master is prevented from entering the slave. When the clock goes to 0 no information can enter the master, but the clock input to the slave is at a 1 and information from the master can now enter into the slave.

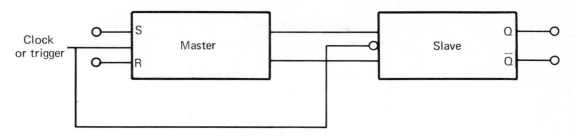

Fig. 7-4. Master-slave clocked FF.

T or toggle FF changes state after a trigger or clock pulse and obeys the truth tables of Table 7-2a. It is common also to indicate the toggle truth table in the manner shown in Table 7-2b. The notation of Table 7-2b means that before the trigger pulse at t_n the bistable is in a state defined as Q, and after one additional trigger pulse, t_{n+1}, the bistable has toggled and is in the state $\overline{Q}$. A T flip-flop is frequently made by cross connecting the inputs and outputs of a master-slave FF, that is, by connecting S to Q and R to Q. Figure 7-5 gives the logic symbol for a T flip-flop.

Table 7-2a. Toggle States.

Before trigger t_n		After trigger t_{n+1}	
Q	Q	Q	$\overline{Q}$
1	0	0	1
0	1	1	0

Table 7-2b. Toggle States.

t_n	Q
t_{n+1}	$\overline{Q}$

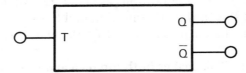

Fig. 7-5. T or toggle FF.

J-K flip-flop is a combination of R-S and T flip-flops. It has two inputs, J and K, the equivalent of the R and S inputs. Whereas in the R-S flip-flop energizing the R-S inputs is forbidden, in the J-K flip-flop energizing the J and K inputs makes the FF toggle. The truth table is given in Table 7-3. Note that the Q values are omitted since they are the opposite of the Q input.

73

Table 7-3. J-K States.

t_n		t_{n+1}
J	K	Q
0	0	Q_n
1	0	1
0	1	0
1	1	$\overline{Q}_n$

IC NAND GATE FFs

The R-S Flip-Flop

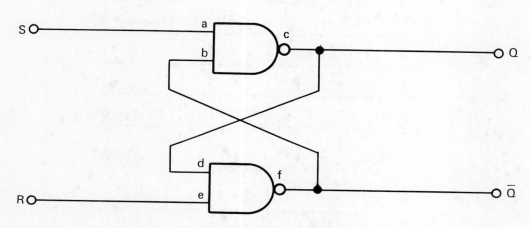

Fig. 7-6. NAND gate R-S flip-flop.

Consider figure 7-6. Define the set conditions as Q = 1, $\overline{Q}$ = 0; define the reset conditions as Q = 0, $\overline{Q}$ = 1. Assume that the FF is in the set condition and S = R = 1. This makes a = 1, b = 0, d = 1, and e = 1. These inputs make output c = 1 and f = 0. This is a stable condition and nothing changes.

Now make S = 0 and R = 1. This makes both inputs a and b = 0, keeping c at 1. Inputs d and e are both 1 keeping f = 0. There is no change. S = 0 and R = 1 are the input conditions to set the FF.

Now make S = 1 and R = 0. With R = 0, input e becomes 0 and output f goes to a 1. a and b are both 1s, making c = 0. Hence S = 1 and R = 0 makes Q = 0, $\overline{Q}$ = 1 and resets the FF.

If both S and R = 0, Q and $\overline{Q}$ go to 1, which is not allowed.

The truth table is given in Table 7-4.

Table 7-4. R-S flip-flop truth table.

S	R	Q	$\overline{Q}$
1	1	No change	
0	1	1	0
1	0	0	1
0	0	Not allowed	

An IC NOR gate FF can be made in the same way. In the NOR gate IC, S = R = 1 is *not allowed*, and S = R = 0 is the stable, *no-change* operating condition.

The Clocked FF

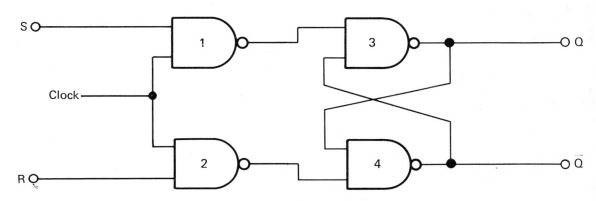

Fig. 7-7. NAND gate clocked FF.

In figure 7-7, gates 3 and 4 form a simple latch. Information is clocked through gates 1 and 2. If the clock is low the outputs of gates 1 and 2 are high. The latch remains in its previous state no matter what the levels of S and R are. Suppose the clock goes high. If S = 1 and R = 0, the output of gate 1 is 0 and of gate 2 is 1. This makes the output of gate 3 a 1 and the output of gate 4 a 0. The truth table is given in Table 7-5.

Table 7-5. Truth Table: NAND gate clocked FF.

Clock	S	R	Q	$\overline{Q}$
1	1	0	1	0
1	0	1	0	1
0	0	1	Does not change Stores last data	
0	1	0		

The D data FF, shown in figure 7-8, is a modification of the clocked FF. Data is available at the D terminal as a 1 or a 0. The inverter ensures that two 0s cannot be presented simultaneously to the information input to gates 1 and 2. Only after the clock pulse has taken place is the data stored and available at Q and $\bar{Q}$.

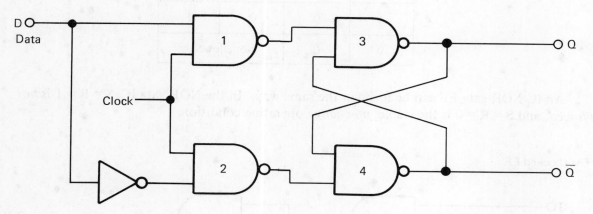

Fig. 7-8. D (data) FF.

Master-Slave clocked FF

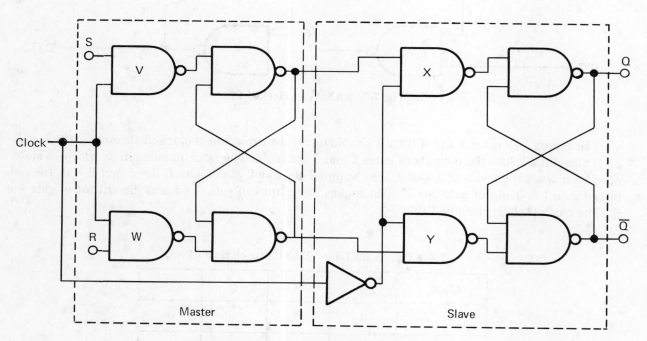

Fig. 7-9. NAND gate master-slave clocked FF.

The master-slave FF (Fig. 7-9) consists of two clocked FFs. When the clock is 1, gates V and W are enabled and information is transferred from S and R to the output of the master. However, the clock input to gates X and Y is at 0 (inputs to X and Y through an inverter), and the master information is blocked from the slave. When the clock goes to 0, the clock input to gates X and Y goes to 1 and the master information is transferred to the slave. The truth table is given in Table 7-6.

Table 7-6. Truth Table: master-slave FF.

S	R	Q	$\overline{Q}$
1	0	1	0
0	1	0	1

Toggle FF from Master-Slave FF

Cross connect S to $\overline{Q}$, R to Q. Assume the clock is a 0, Q = 1, and $\overline{Q}$ = 0. The cross connection makes S = 0, R = 1. With the clock low nothing happens. Now let the clock go to 1 and back to 0. The truth table of the previous section tells us that if S = 0 and R = 1 the output must go to Q = 0, $\overline{Q}$ = 1 after the clock has gone to 1 and back to 0. But this has changed the state of the FF, which is a toggle. Similarly, each clock pulse causes the FF to toggle.

J-K Flip-Flop

The J-K flip-flop (Fig. 7-10) is a modification of the toggle FF. Input gates V and W are modified to have three inputs. The additional inputs are called the J-K inputs.

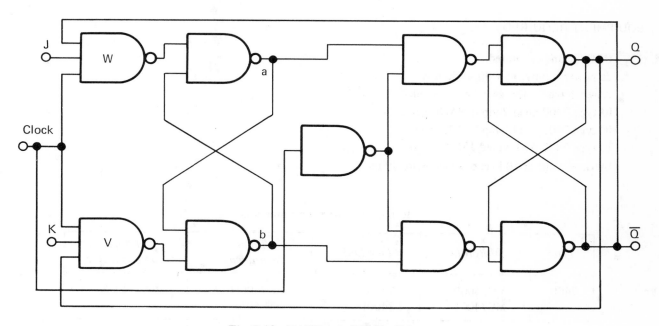

Fig. 7-10. NAND gate J-K flip-flop.

If J and K are both 1 the FF operates as a toggle FF as in the previous section. This must be so, since in Boolean logic 1·A = A.

Now, let J = 1 and K = 0. If $\overline{Q}$ = 1 and Q = 0 when the clock goes to 1, then there are three 1 inputs to gate W. The output of gate W goes to 0. The output of gate V goes to 1. This makes the output of the master at point a = 1 and at point b = 0. When the clock goes to 0, the information at a and b is transferred to the slave, making Q = 1 and $\overline{Q}$ = 0.

Again, let J = 1 and K = 0. Suppose Q = 1 and $\overline{Q}$ = 0. The only way Q could have become a 1 is for a to have been a 1 and b a 0. When the clock goes to 1, what are the inputs to gates W and V? Gate W: clock = 1, J = 1, $\overline{Q}$ = 0. Gate V: clock = 1, J = 0, $\overline{Q}$ = 1. Hence the outputs of gates W and V are 1 and the master latch does not change, and in this case also Q = 1 and $\overline{Q}$ = 0.

Similar arguments hold for J = 0, K = 1. The truth table for the J-K flip-flop is shown in Table 7-7. The truth table for this FF shows that it is an RST FF with the toggle properties occurring when both J = K = 1.

Table 7-7. Truth table: J-K NAND gate FF.

t_n Before clock pulse		t_{n+1} After clock pulse		
J	K	Q	$\overline{Q}$	
1	0	1	0	
0	1	0	1	
1	1	$\overline{Q}$	Q	Toggle
0	0	Q	$\overline{Q}$	No change

EQUIPMENT REQUIRED

CRO, dc and calibrated.

dc power supply, +5 volts at 50 mA.

2 switch banks, five switches per bank.

IC type 7400 quad 2-input NAND gate.

IC type 7402 quad 2-input NOR gate.

IC type 7472 AND-gated J-K flip-flop.

SWG, +5 volts at 10 kHz or single-pulse 50 μs.

IC manufacturers' part numbers.

Type	Motorola	Fairchild	Texas Instruments	National Semiconductor
7400	MC7400P MC7400L	7400PC 7400DC	SN7400N SN7400J	DM7400N
7402	MC7402P MC7402L	7402PC 7402DC	SN7402N SN7402J	DM7402N
7472	MC7472P MC7472L	7472PC 7472DC	SN7472N SN7472J	DM7472N

For all ICs in this experiment: V_{cc} = +5 Volts to pin 14, 0 (ground) to pin 7.

In each part of the experiment take CRO voltage measurements at the points indicated to complete the accompanying tables. The CRO vertical amplifier is to be used dc coupled with a gain setting of 1 volt/div and is to be used to make all the voltage measurements. Measure all voltages to within *0.1 volt*. *All data is to be taken row by row.*

In each part of this experiment measurements have to be made at the two outputs Q and $\bar{Q}$ of the FF. If a dual-trace or dual-beam CRO is available, both traces should be used for the two points. If a single-trace CRO is used the measurements will be made more rapidly and the FF operation seen more readily if the CRO is connected to the Q and $\bar{Q}$ outputs by means of a switch. In all measurements connect the Q side to the same CRO input terminal or to the same side of the switch.

1. R-S Flip-Flop with NAND Gates

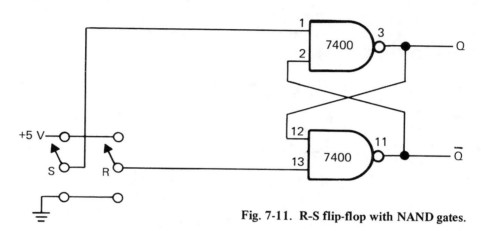

Fig. 7-11. R-S flip-flop with NAND gates.

Table 7-8E. NAND gate R-S flip-flop.

	S	*R*	*Q*	*$\bar{Q}$*
1	+5	0		
2	+5	+5		
3	0	+5		
4	+5	+5		
5	0	0		
6[a]	+5	+5		
7	0	0		
8[b]	+5	+5		

[a]Switch S and R to +5 almost simultaneously but switch S is to reach +5 before switch R. This is to simulate a race condition.

[b]Switch S and R to +5 almost simultaneously but switch R is to reach +5 before switch S. Again, this is to simulate a race condition.

2. Gated R-S Flip-Flop.

(a) Gated R-S Flip-Flop with NAND Gates

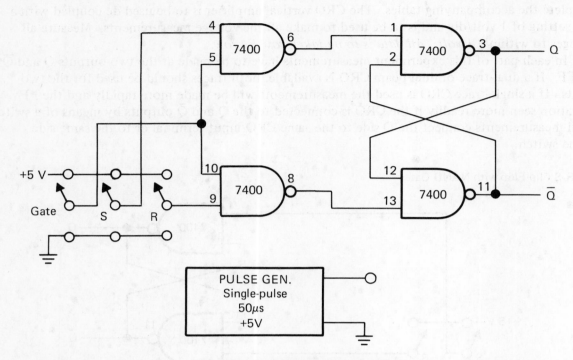

Fig. 7-12. Gated FF with NAND gates.

Table 7-9Ea. Gated FF with NAND gates.

	Gate	S	R	Q	$\bar{Q}$
1	+5	0	+5		
2	+5	+5	0		
3	+5	0	+5		
4	+5	0	+5		
5	0	+5	0		
6	0	0	+5		
7	0	0	0		
8	0	+5	+5		
9	0	0	+5		

80

In figure 7-12, disconnect pins 5 and 10 from the gate switch and connect pins 5 and 10 to the single pulse generator.

For each row, set switches S and R to the indicated values and read the values of Q and $\overline{Q}$. Then apply the single pulse where required and again read the values of Q and $\overline{Q}$. Note: The "Before Single Pulse" data for each row is the "After Single Pulse" data of the previous row.

Table 7-9Eb. Gated FF with NAND Gates.

			Before Single Pulse		After Single Pulse	
	S	*R*	*Q*	*$\overline{Q}$*	*Q*	*$\overline{Q}$*
1[a]	+5	0	–	–		
2	0	+5				
3	+5	0				
4	0	+5				
5	0	0				
6	+5	0				
7	0	+5			No	
8	+5	0			Single	
9	0	+5			Pulse	
10	5	0				
11	0	5				

[a]This sets the FF in a known state.

3. The D FF

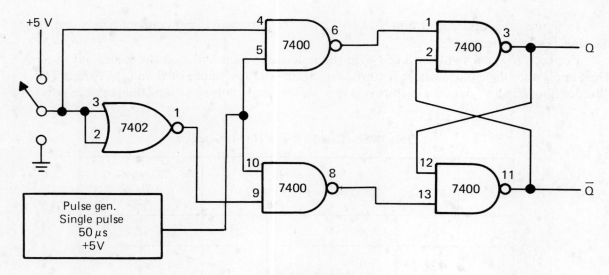

Fig. 7-13. D FF.

As in part 2b, the "Before Clock Pulse" entries are obtained from the "After Clock Pulse" data from the previous row.

Table 7-10E. DFF

	Info	Before Clock Pulse		After Clock Pulse	
		Q	$\overline{Q}$	Q	$\overline{Q}$
1[a]	+5	–	–		
2	0				
3	+5				
4	0				
5	0				
6	+5			No	
7	0			Clock	
8	+5			Pulse	
9	0				
10	5				

[a]This sets the FF in a known state.

82

4. AND gated J-K master-slave FF

   (a)   AND gated J-K operation

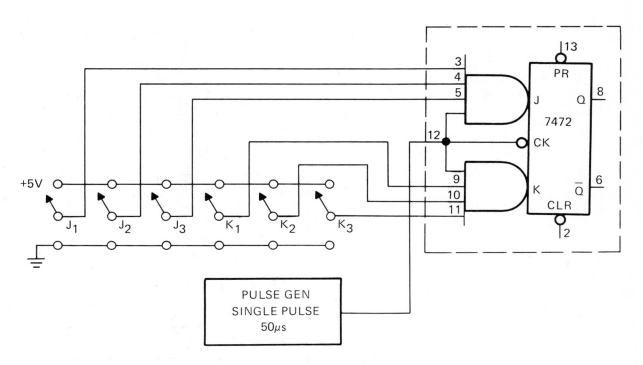

Fig. 7-14a. AND gated J-K FF.

In Tables 7-11Ea and 7-11Eb, the notation t_n refers to the state of the FF at the time t_n before the clocking pulse. t_{n+1} refers to the state of the FF at time t_{n+1} after the clocking pulse.

Table 7-13Ea. AND gated J-K operation

	J_1	J_2	J_3	K_1	K_2	K_3	Before Single Pulse t_n		After Single Pulse t_{n+1}	
							Q	$\overline{Q}$	Q	$\overline{Q}$
a	0	0	0	+5	+5	+5	–	–		
1	0	0	0	0	0	0				
2	+5	0	0	0	0	0				
3	+5	+5	0	0	0	0				
4	+5	+5	+5	0	0	0				
5	0	0	0	+5	0	0				
6	0	0	0	+5	+5	0				
7	0	0	0	+5	+5	+5				
8	+5	+5	0	+5	+5	0				
9	+5	+5	+5	+5	+5	0				
10	+5	+5	0	+5	+5	0				
11	+5	+5	0	+5	+5	+5				
12	+5	+5	+5	+5	+5	+5				
13	+5	+5	+5	+5	+5	+5				
14	+5	+5	+5	+5	+5	+5				

aTo clear the FF

(b)   J–K Mode Using 10–kHz SWG

Change the output of the pulse generator to 10 kHz square wave. Set the CRO controls to:

AUTO trigger

NEG slope

INT trigger

View the output of the SWG on the CRO. Adjust the horizontal time/div so that the waveform shown in figure 7-14b. is obtained. View and record the waveforms on the graph paper of figure 7-14b for the following conditions. *DO NOT* change the time/div scale. **Trigger CRO as indicated.**

84

Part	Output	Trigger	J_1	J_2	J_3	K_1	K_2	K_3
4(ba)	Q	EXT from Q	+5	+5	+5	+5	+5	+5
4(bb)	$\overline{Q}$	EXT from Q	+5	+5	+5	+5	+5	+5
4(bc)	Q	INT	+5	+5	+5	0	+5	+5
4(bd)	$\overline{Q}$	INT	+5	+5	+5	0	+5	+5
4(bc)	Q	INT	0	+5	+5	+5	+5	+5
4(bf)	$\overline{Q}$	INT	0	+5	+5	+5	+5	+5

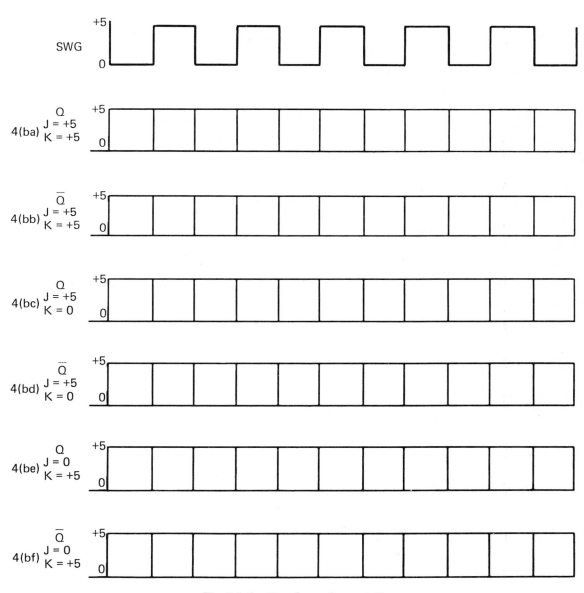

Fig. 7-14b. Waveforms for part 4b.

(c) Asynchronous Mode.

PRESET (PR) and CLEAR (CLR).

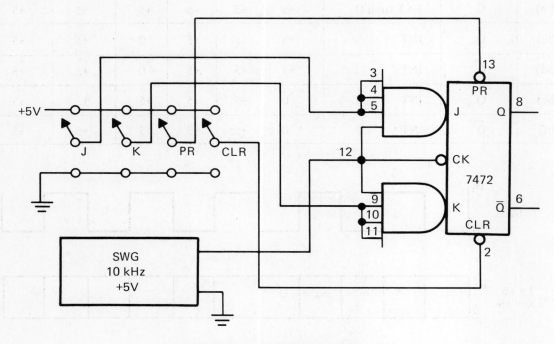

Fig. 7-14c. Asynchronous mode.

The PR input is used to set the Q output to a 1, and the $\overline{Q}$ output to a 0. The CLR input is used to set the Q output to a 0 and the $\overline{Q}$ output to a 1. The PR and CLR inputs override all other inputs.

In Table 7-11Eb, tabulate the voltages obtained for the following conditions. If a square wave is seen, use the notation SW.

Table 7-11Eb. Asynchronous operation.

	J	K	PR	CLR	Q	$\overline{Q}$
1	+5	+5	+5	+5		
2	+5	+5	0	+5		
3	+5	0	0	+5		
4	0	+5	0	+5		
5	0	0	0	+5		
6	+5	+5	+5	0		
7	+5	0	+5	0		
8	0	+5	0	+5		
9	0	0	0	+5		
10	+5	+5	0	0		
11	+5	0	0	0		
12	0	+5	0	0		
13	0	0	0	0		

REQUIRED RESULTS

In completing the following tables use positive logic, $1 \equiv\, > 2.5$ volts, $0 \equiv\, < 0.5$ volt.

5. R-S Flip-Flop.

Use the following notation in completing the truth tables:

1 and 0 for the logic states
NC for no change
NA for not allowed

Based on the data of Table 7-8E, complete Table 7-8R, the truth table for the R–S flip-flop using NAND gates.

Table 7-8R. R-S flip-flop with NAND gates.

S	R	Q	$\overline{Q}$
1	0		
0	1		
1	1		
0	0		

6.

Based on the data of Tables 7-9Ea and 7-9Eb and using the same notation as in part 5, complete the following truth table, Table 7-9R, for the gated R-S flip-flop of figure 7-12. The Q and $\overline{Q}$ columns are the output levels after the gating pulse.

Table 7-9R. Gated NAND gate R-S flip-flop.

S	R	Q	$\overline{Q}$
1	0		
0	1		
0	0	NC	NC
1	1	NA	NA

7.

Based on the data of Table 7-10E and using the same notation as in parts 5 and 6, complete Table 7-10R for the D flip-flop.

Table 7-12R. D flip-flop.

Info	t_n		t_{n+1}	
	Q	$\overline{Q}$	Q	$\overline{Q}$
1	1	0		
0	1	0		
1	0	1		
0	0	1		

8. AND gated J-K FF.

In completing the tables for this FF use the following notation:

1,0 logic levels
X — state of this input does not affect the state of the output
NC — no change

(a) Clocked J-K mode.

Use the results of Table 7-11Ea and waveforms of part 4(b) to complete Table 7-13Ra. Note that the FF is symmetrical; the effect of the J inputs on Q is identical to the effect of the K inputs on $\overline{Q}$.

Table 7-11Ra. J-K Mode

J_1	J_2	J_3	K_1	K_2	K_3	t_n		t_{n+1}	
						Q	$\overline{Q}$	Q	$\overline{Q}$
1	1	1	0			—	—		
1	1	1		0		—	—		
1	1	1			0	—	—		
0			1	1	1	—	—		
	0		1	1	1	—	—		
		0	1	1	1	—	—		
0			0			1	0		
0			0			0	1		
1	1	1	1	1	1	1	0		
1	1	1	1	1	1	0	1		

(b)

Using the data of Table 7-11Eb of part 4, complete Table 7-11Rb for the asynchronous mode. Indicate toggling by $\overline{Q}$ in the Q column and Q in the $\overline{Q}$ column.

Table 7-11Rb. Asynchronous Mode.

PR	CLR	Q	$\overline{Q}$
1	1		
0	1		
1	0		
0	0		

DISCUSSION

1. Explain how the 2-input NAND gate FF of part 1 and figure 7-11 has bistable properties and will stay in the stable state. Explain how the S&R inputs change the state of the bistable.

2. Explain why, in figure 7-11, S = 1, R = 1 is allowed but S = 0, R = 0 is not allowed.

3. An R-S bistable has momentarily been driven into the nonallowed state. Based on the data of lines 6 and 8 of Table 7-8E, discuss the factors which determine into which of the allowed states it may return.

4. Explain why, in Table 7-9Ea, the results of lines 1, 2, and 3 differ from those of lines 4, 5, 6, 7, 8, and 9.

5. It is desired to use a latch for temporary data storage of a completed arithmetic operation in a computer while another arithmetic operation is taking place. Based upon the data of Table 7-9Ea, what gate level, 1 or 0, should be used to transfer data into the latch? What level should be used to isolate the latch from the occurring arithmetic operation?

6. Based on the data of Table 7-9Eb, what does the single gate or strobe pulse do?

7. The gated R-S flip-flops of part 2 have four gates. The following diagram, figure 7-15, is an attempt to make a gated FF with two gates. However, it will not work correctly. Explain why it will not. (Hint: Consider all possible levels for R, S and the gate.)

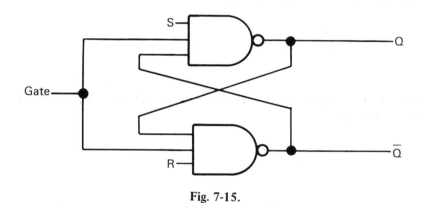

Fig. 7-15.

8. The standard method of writing a truth table for the type 7472 is given in Table 7-14D.

Table 7-14D.

Function table.

INPUTS		OUTPUTS				
PRESET	CLEAR	CLOCK	J	K	Q	$\overline{Q}$
L	H	X	X	X	H	L
H	L	X	X	X	L	H
L	L	X	X	X	H*	H*
H	H	⊓	L	L	Q_0	$\overline{Q}_0$
H	H	⊓	H	L	H	L
H	H	⊓	L	H	L	H
H	H	⊓	H	H	TOGGLE	

positive logic: J = J1 · J2 · J3; K1· K2 · K3
Source: Texas Instruments Inc.

(a) Compare this with Table 7-11Ra and the waveforms of part 4b and explain why the J-K entries have the same meaning in both tables.

(b) To obtain a 1 in the Q column requires that J_1, J_2 and J_3 be at 1. To obtain a 0 in the Q column requires that K_1, K_2 and K_3 be at 1. How is this requirement shown on the type 7472 logic diagram?

(c) In the OUTPUTS column, what is the significance of the asterisks (*) next to the Hs in the third row?

(d) In the fourth row, what is the meaning of the Q_0 and $\overline{Q}_0$ entries?

9. From the wave pattern for part 4(b), what is the frequency of the output at Q? Show your calculations.

10. What is the frequency that would be obtained at the Q output of FF 2?

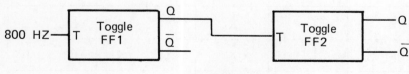

Fig. 7-16.

11. The logic diagram of the type 7472 has an inverting circle at the clock input. This means that data is transferred to the slave output when the clock goes to its logic zero level. The PR and CLR inputs have inverting circles also. What do these mean? (Hint: See Table 7-11Eb and Table 7-12d.)

OBJECT

(a)   To study the operation of binary counters.

(b)   To study binary counting and the representation of numbers in the binary number system.

INTRODUCTORY THEORY

THE BINARY COUNTER

A binary or bistable FF has the characteristic of changing state with each incoming pulse and reverting to its original state after two pulses have been applied. After the second pulse there is nothing to indicate whether zero or two pulses have occurred. However, if the return to the original state is considered an "overflow" and is applied to a second binary, it now also will change its state; this change in state combined with that of the first binary can be used to indicate that two pulses have occurred.

Additional binaries, similarly, can be cascaded beyond the first two and their states will indicate the number of pulses that have been applied to the input of the succession of binaries called "binary counter." In such a counter the count is represented in binary notation by the states of the counter binaries or stages.

If we represent the state of the binary before a pulse has appeared as a 0, and the opposite state as a 1, we can then represent the number of incoming pulses in both binary notation and by the state of the counter stages in an n-stage counter as

$$\text{count} = B_n \, 2^{n-1} + B_{n-1} \, 2^{n-2} + B_{n-2} \, 2^{n-3} + \cdots + B_1 \, 2^0 \tag{8-1}$$

In this representation the B_S (binary digits or bits) can be either 0 or 1. For example, suppose we have a four-stage counter and 12 pulses have been applied. After 12 pulses we would find the binary states (and Q outputs) as follows:

stage	level
4	1
3	1
2	0
1	0

And the count from Eq. (8-1) is

$$\text{count} = 1 \times 2^{(4-1)} + 1 \times 2^{(3-1)} + 0 \times 2^{(2-1)} + 0 \times 2^{(1-1)} = 12$$

$$= 1 \times 2^3 + 1 \times 2^2 + 0 \times 2^1 + 0 \times 2^0 = 12$$

In a binary, both outputs, the Q and $\overline{Q}$ terminals, can be made available, or only the single terminal Q can be made available. When only the single terminal is available, the state of the various stages are represented by the Q levels—either 1 or 0—and these levels are the B values in Eq. (8-1). If both sides of the binary are available, the counter becomes more universal and can be used for multiple coincidence outputs or decoding; the complements of the count are available at the $\overline{Q}$ terminals. Thus the $\overline{Q}$ terminals count *down*, as opposed to the Q terminals which count up.

A binary counter must have provision for *clearing* the count so that it is at a total count of 0 before any new count pulses are applied to the counter if correct counts are to be obtained each time a new train of pulses is to be counted.

Frequency Division

When the incoming pulses occur at a regular rate or constant frequency, the frequency at the output of the first binary is one-half that of the incoming frequency. Each binary in turn divides its incoming frequency by 2 and, therefore, in an n-stage counter the output frequency at any stage is $\frac{1}{2}^{p} \times$ the incoming frequency, where p is the number of stages up to the point of measurement.

The Ripple Counter

Figure 8–1a shows in block diagram a basic type of binary counter called a ripple counter. In this type of counter the output of each FF triggers the next stage. This triggering takes place on a 1

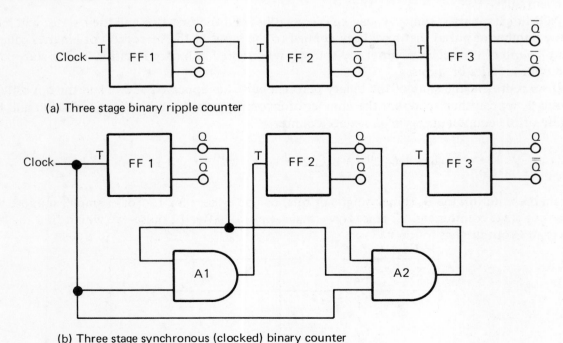

(a) Three stage binary ripple counter

(b) Three stage synchronous (clocked) binary counter

Fig. 8-1. Cascade binary counters.

to 0 transition. Suppose the condition exists in which all three FFs are in the 1 state and a trigger pulse arrives at the input to FF 1. Any circuit has some propagation or switching time, and FFs are no exception. As a result of this propagation or delay time, each stage arrives at its final state at a time delayed by the previous stages, and the count "ripples" through the counter. For example, if the switching time per bistable were 100 ns, in a three-stage counter such as shown in figure 8-1a, the output of FF 3 would arrive at its final state 300 ns after the trigger pulse, if all three FFs had to change state.

From a counting standpoint or from a frequency-division standpoint this causes no problems, but consider the following problem. Suppose we are supplying input pulses to a binary counter at a 200 ns rate and we have an operation which must be performed at the count where stages 1 and 5 in a five-stage counter are at a 1. This requires a coincidence or AND gate. But the fifth stage does not arrive at its correct state until 500 ns of delay and at this time two additional trigger pulses (or counts) have been applied to the counter. The AND gate, therefore, is operated at an incorrect count because of the FF switching times. Other coincidence gates may similarly give false indications or may be turned on and off at wrong times and thereby provide false triggering pulses.

The Synchronous (Clocked) Binary Counter

The above difficulties with the ripple counter may be overcome with a counter in which all stages are triggered simultaneously. The counter becomes more complex and is shown in block form in figure 8-1b. Examination of figure 8-1b shows that the trigger to FF 2 comes from AND gate A1. Gate A1 is enabled by a 1 from the Q output of FF 1 and the clock pulse. Except for the gate A1 delay (which in general is low compared to that of a FF), both FF 1 and FF 2 are triggered simultaneously by the clock. The same is true for FF 3, which is clock triggered when FF1 and FF 2 are in the 1 state. Hence, except for the gate delays, all stages are triggered simultaneously.

CRO Waveforms and Triggering

The observation of counter wave patterns on a cathode-ray oscilloscope (CRO) provide a powerful and rapid method of analysis of counter operation and possible malfunction. To obtain the correct pattern and analyze it requires that the CRO be triggered at the correct time on the waveform. Consider the wave patterns of a three-stage binary counter triggered on the 1 to 0 transition (positive logic) as shown in figure 8-2. To obtain the correct patterns, note that all the patterns start with a negative-going transition, The only time when all three FFs simultaneously have a negative-going transition is at the time when the Q output of FF 3 is going negative. Hence,

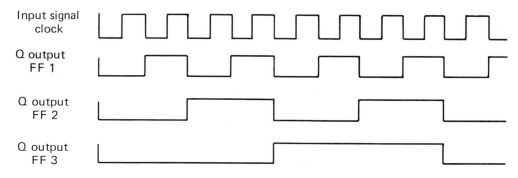

Fig. 8-2. Wave patterns for the three-stage binary counter of Fig. 8-1.

to properly trigger the CRO we must use external, negative-slope triggering from the Q output of FF 3.

EQUIPMENT REQUIRED

CRO, dc and calibrated.

dc power supply, +5 volts at 100 mA.

SWG +5 volts at 10 kHz or Single-Pulse 50 μs.

IC type 7400 quad 2-input NAND gate.

IC type 7420 dual 4-input NAND gate.

IC type 7472 AND gated J–K flip-flop. Three required.

0.1 μf capacitor.

IC manufacturers' part numbers.

Type	Motorola	Fairchild	Texas Instruments	National Semiconductor
7400	MC7400P MC7400L	7400PC 7400DC	SN7400N SN7400J	DM7400N
7420	MC7420P MC7420L	7420PC 7420DC	SN7420N SN7420J	DM7420N
7472	MC7472P MC7472L	7472PC 7472DC	SN7472N SN7472J	DM7472N

EXPERIMENTAL PROCEDURE

For all ICs in this experiment: V_{cc} = +5 volts to pin 14, 0 (ground) to pin 7.

1.

(a) Wire the following circuit of figure 8-3 for a three-stage binary ripple counter.

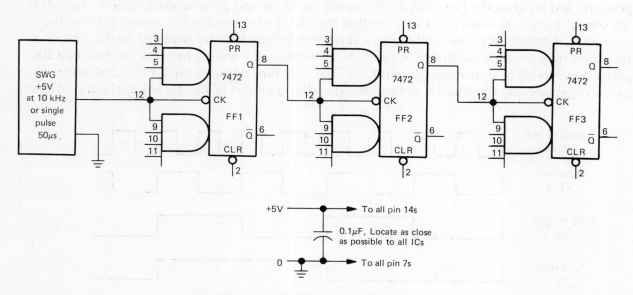

Fig. 8-3a. IC—Binary ripple counter.

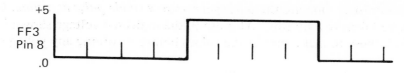

Fig. 8-3b. FF 3-pin 8-wave pattern.

View the wave shape at the output of the pulse generator. Use AUTO triggering, negative slope, and INT. Vertical is dc coupled 1 volt/div. Either vary the pulse generator frequency or adjust the CRO sweep speed so that 10 cycles of the pulse generator occupy *exactly* 10 horizontal divisions. Now view the wave pattern at pin 8 of FF 3. It should have the appearance shown in figure 8-3b. **DO NOT** at any time during the experiment change the pulse generator frequency or the CRO horizontal sweep frequency.

(b)   Change the pulse generator function to SINGLE PULSE.

Check the voltage levels at pin 8 of all three FFs. At least one should be at ground and at least one should be at approximately +5 volts. If this condition does not exist, return the pulse generator momentarily to SQUARE WAVE and then back to SINGLE PULSE and once again check the pin-8 voltage levels. If necessary repeat again until the pin-8 levels are not all alike.

Measure the voltage at the indicated IC terminal pins to complete Table 8-1Ea.

Table 8-1Ea.

FF 3		FF 2		FF 1	
Pin 8 Q	Pin 6 $\overline{Q}$	Pin 8 Q	Pin 6 $\overline{Q}$	Pin 8 Q	Pin 6 $\overline{Q}$

(c)   This can be done in the order indicated, or all pin 2s can be connected to ground simultaneously.

1.   Connect pin 2 of FF 1 to ground and then remove the pin 2 ground connection.
2.   Connect pin 2 of FF 2 to ground and then remove the pin 2 ground connection.
3.   Connect pin 2 of FF 3 to ground and then remove the pin 2 ground connection.

Measure the voltage at the indicated IC terminal pins to complete Table 8-1Eb.

Table 8-1Eb.

FF 3		FF 2		FF 1	
Pin 8 Q	Pin 6 $\overline{Q}$	Pin 8 Q	Pin 6 $\overline{Q}$	Pin 8 Q	Pin 6 $\overline{Q}$

(d)   Binary counting. Operate the pulse generator, *a single pulse at a time*, to complete Table 8-1Ec. Data is to be taken row by row. At pulse 0, the indicated voltage shown in Table 8-1Ec should be obtained. Check these to make sure of the values before recording any voltages.

Table 8-1Ec.

Pulse	FF 3		FF 2		FF 1	
	Pin 8 Q	Pin 6 $\overline{Q}$	Pin 8 Q	Pin 6 $\overline{Q}$	Pin 8 Q	Pin 6 $\overline{Q}$
0	0	$\approx +5$	0	$\approx +5$	0	$\approx +5$
1						
2						
3						
4						
5						
6						
7						

(e)   CRO wave patterns—correct triggering. Set pulse generator to 10 kHz SW. View the wave forms at the IC terminal pins indicated. Sketch them as indicated on the graph paper of figure 8-3c. For the CRO triggering use AUTO trigger, negative slope, dc coupled and *external* triggering from pin 8 (Q output) of FF 3. Vertical input, dc coupled, 1 volt/div.

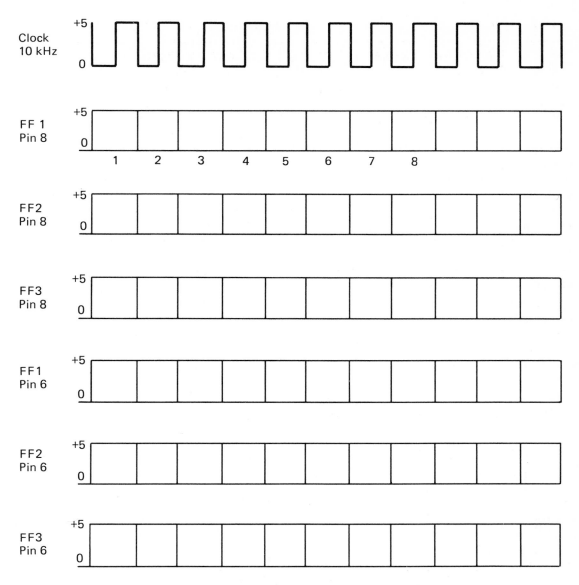

Fig. 8-3c. Wave patterns for part 1e. Correct triggering.

2. CRO Wave Patterns—Incorrect Triggering

The patterns obtained in part 1(e) give the correct wave pattern. This part of the experiment will demonstrate how incorrect patterns can be obtained with improper triggering. In describing the patterns, indicate the description by means of the following letter code. If necessary, use more than one letter to describe your results. Take the data *column by column* starting with column A.

(a) Correct.
(b) Wrong phase—correct frequency.
(c) Unsteady or unrecognizable.

Table 8-2E.

Wave Shape at Pin	EXT NEG TRIGGER from clock	EXT NEG TRIGGER from FF 1, Pin 8	EXT NEG TRIGGER from FF 2, Pin 8	EXT NEG TRIGGER from FF 3, Pin 6	EXT POS TRIGGER from FF 3, Pin 6	NEG TRIGGER INT	POS TRIGGER INT
	A	B	C	D	E	F	G
Clock							
FF 1, Pin 8							
FF 1, Pin 6							
FF 2, Pin 8							
FF 2, Pin 6							
FF 3, Pin 8							
FF 3, Pin 6							

3.

   (a)   Synchronous (CLOCKED) Counter—Three-Stage

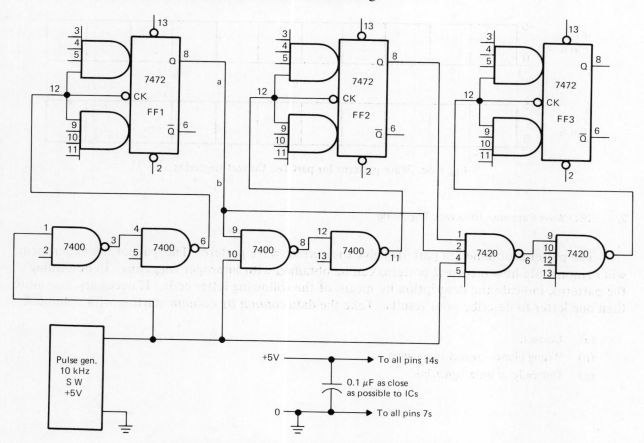

Fig. 8-4a. Three-Stage synchronous (clocked) binary counter.

Wire the circuit shown in figure 8-4a. This is a three-stage clocked binary counter. The wave shapes obtained at pins 8 and 6 should be identical to those obtained in part 1e of the experiment. If they are not obtained, temporarily change the synchronization to INT and make sure that each stage, starting from FF 1, operates properly. When all stages operate properly, sketch the wave shapes on figure 8-4b at the points required. (Use external synchronization—negative slope—from pin 8 of FF 3.)

(b) Synchronous Counter—Effect of Wiring Error. In wiring a complex circuit a wiring error occasionally occurs, or a component may be defective. Analysis of the wave shapes obtained provides a clue to the location of the error. Even in a simple circuit such as that of figure 8-4a, the possibilities of a fault are many. In this part of the experiment a fault is introduced. An analysis of the wave shapes obtained with the correct ones of part a should determine the cause of the fault.

The fault to be introduced is: "Remove the connection ab from pin 8 of FF 1 which goes to pin 9 of the type 7400 IC and pin 2 of the type 7420 IC." Now sketch the wave shapes on figure 8-4c at the points indicated.

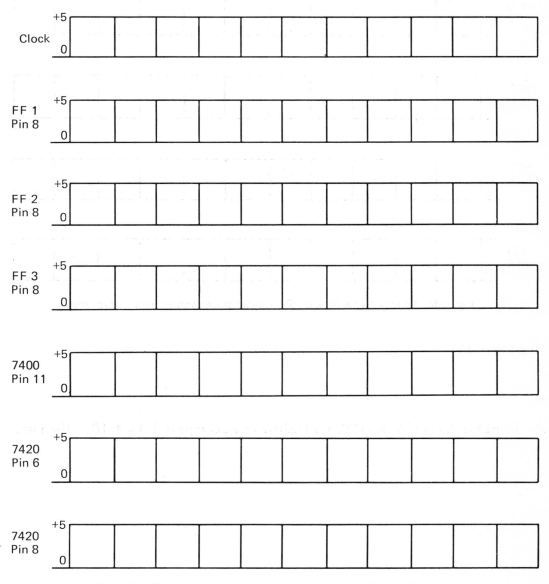

Fig. 8-4b. Wave patterns for part 3a—synchronous (clocked) counter.

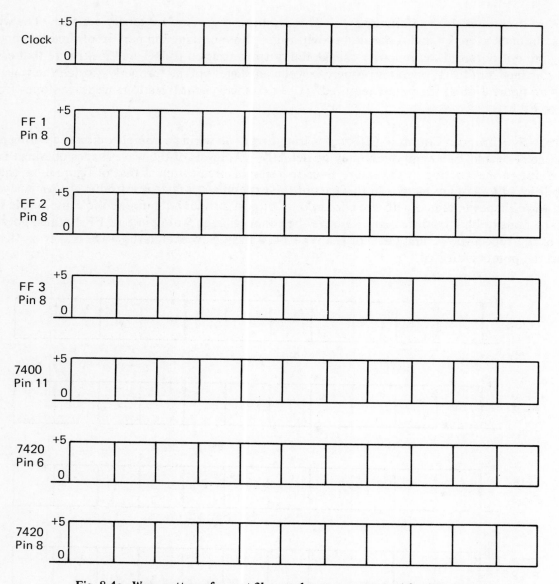

Fig. 8-4c. Wave patterns for part 3b—synchronous counter with wiring error.

REQUIRED RESULTS

4.

   (a)   Using the data of Table 8-1Ec for positive logic, complete Table 8-1Rc. Use logic levels 1 and 0.

Table 8-1Rc.

Pulse	Q_3 Pin 6 FF 3	Q_2 Pin 6 FF 2	Q_1 Pin 6 FF 1	Binary Number $Q_3Q_2Q_1$	Decimal Equiv. of K	$\overline{Q}_3$ Pin 9 FF 3	$\overline{Q}_2$ Pin 9 FF 2	$\overline{Q}_1$ Pin 9 FF 1	Binary Number $\overline{Q}_3\overline{Q}_2\overline{Q}_1$	Decimal Equiv. of L	Col. A + Col. B
				K	A				L	B	
0	0	0	0	0 0 0	0	1	1	1	1 1 1		
1											
2											
3											
4											
5											
6											
7											

(b) Using the wave patterns of part 1(e), for positive logic and at the indicated horizontal box, complete Table 8-1Re using logic levels 1 or 0 as obtained.

Table 8-1Re.

Box	Q_3	Q_2	Q_1	$\overline{Q}_3$	$\overline{Q}_2$	$\overline{Q}_1$
1						
2						
3						
4						
5						
6						
7						
8						

1. Explain the function and necessity for the grounding of the pin 2s of each of the FFs in part part 1(c). What does grounding all the pin 2s do? What was the binary count before and after the pin 2s were grounded?

2. Referring to Table 8-1Rc, column A is the _____ complement of column _____.

3. Compare the progression of the count in column A and column B of Table 8-1Rc. Discuss with respect to both binary and decimal representation.

4. What is the maximum count capability of an eight-stage binary counter?

   Binary _____   Decimal _____

5. Compare Table 8-1Rc against Table 8-1Re and discuss. In your discussion, comment on ease of analysis of correct counter operation using the method of part 1(d) or part 1(e). Which method of operation SWG or single pulse would you use if you had only a voltmeter available?

6. If the clock frequency is 10 kHz. calculate the frequency at pin 8 of FF 1, pin 8 of FF 2, and pin 8 of FF 3. Show your calculations based upon a sweep speed of 100 μs/div. per division.

7. Calculate the output frequency of a seven-stage binary ripple counter if the input frequency is 1 MHz.

8. In part 2, which method of triggering (trigger point, trigger slope, column of Table 8-2F) gave the correct wave pattern at all points in the counter. Explain why it gave the correct pattern. [Hint, refer also to part 1(e)].

9. Explain the operation of figure 8-4a with respect to the wave patterns of part 3(a).

10. The wiring fault introduced in part 3(b) caused malfunctioning of the counter. Explain the resultant waveforms at pin 8 of FF 2 and pin 8 of FF 3 based on your analysis of the error and the waveforms obtained in part 3(b).

11. What are the advantages and disadvantages of the ripple versus the synchronous counter?

12. A binary has a switching time delay of 25 ns. It is used as the FF element in a six-stage ripple counter. The counter is in the 111111 state. A pulse changes the first stage to the 0 state. How long after the first stage changes to the 0 state will the sixth stage change to the 0 state?

Experiment 9

DIVIDE-BY-N COUNTERS;
DECADE COUNTERS

OBJECT

    (a)   To study counters which divide the incoming frequency by counts other than binary powers and which use cascaded T flip-flops.

    (b)   To study the count states of these counters.

    (c)   To study the weighted BCD decade counters 8421 and $2'421$.

    (d)   To study the unused states of the 8421 counter.

INTRODUCTORY THEORY

    The toggle or T flip-flop divides an incoming frequency by a factor of 2. Cascading T flip-flops results in a final frequency which is the original frequency divided by $\frac{1}{2}^p$ where p is the number of stages. Many applications require output frequencies which are a division of the input frequency by a factor other than a power of 2. For example, the divide-by-10 counter is very useful and common.

    What is meant by a divide-by-N counter? This means primarily that there is an output frequency which is 1/N of the input frequency. The output wave shape, not necessarily sine wave or square wave, repeats itself at a repetition rate 1/N of the input frequency. A common way to build a counter is to cascade T flip-flops. Let us assume that we have such a counter and that no incoming signal is applied to the counter. We can examine the state of each binary and express it in binary or equivalent notation. Let the signal voltage be applied. For each cycle or pulse of the incoming signal one or more of the FFs changes state. The FFs change state so that each incoming pulse results in a different overall count state. In a divide-by-N counter there are N different count states. After the Nth count the counter has returned to its original state and the process is repeated. Such a counter, with its N different count states, is frequently designated as a counter of *modulus N* or *mod N*.

    The basis for mod-N counters is the cascaded T flip-flop. Since the T flip-flop of p stages has 2^p count states, to divide by less than 2^p requires that some of the count states must be skipped. The number of T flip-flop stages that are required depends upon the count. A mod-N counter requires p stages for values of N which fall between 2^p and 2^{p-1}. For example, a four-stage counter (as a binary counter capable of dividing by 16) is required for values of N between 9 and 15.

    IC mode-N counters utilize the R–S and J–K properties of the master–slave FF to skip the required number of unwanted states. Both ripple and synchronous (clocked) counters can be built, but ripple counters are more simple.

A mod-3 counter which makes use of the J–K properties of the master–slave universal FF is shown in figure 9-1. Since it has two stages it is fundamentally capable of dividing by 4. How does this counter skip one count state and divide by 3?

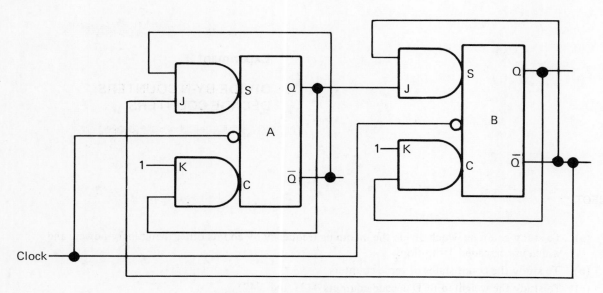

Fig. 9-1. Mod-3 J–K, R–S counter.

In the circuit diagram of figure 9-1, $\overline{Q_B}$ is connected to the J input of FF_A. The K input of FF_A is connected to logic 1. The state of FF_B controls the operation of FF_A. If $\overline{Q_B} = 1$, J_A and K_A are both 1 and FF_A toggles after a clock pulse. If $\overline{Q_B} = 0$, $J_A = 0$ and $K_A = 1$ and FF_A will go to a 0 state ($Q_A = 0$, $\overline{Q_A} = 1$) after a clock pulse. FF_B operates in almost the same manner. J_B is connected to Q_A and K_B to logic 1. If $Q_A = 1$, FF_B toggles (J_B and $K_B = 1$). If $Q_A = 0$, ($J_B = 0$, $K_B = 1$) FF_B goes to the 0 state ($Q_B = 0$, $\overline{Q_B} = 1$) after a clock pulse.

To start, assume that both FFs are in the 0 state.

Count	J_A	K_A	Q_A	$\overline{Q_A}$	J_B	K_B	Q_B	$\overline{Q_B}$	
0			0	1			0	1	
	1	1	0	1	0	1	0	1	
1			1	0			0	1	FF_A has toggled FF_B forced to stay the same
	1	1	1	0	1	1	0	1	
2			0	1			1	0	Both FF's toggle since $J_A = K_A = 1$ and $J_B = K_B = 1$
	0	1	0	1	0	1	1	0	
			0	1			0	1	FF_A forced to stay the same FF_B forced into the 0 state

The counter is now back to its original state after three pulses and after three distinct count states.

The most common mod-N counter is the divide-by-10 or the decade counter. It requires four binary stages. Counters can be designed to have any possible state of the four binaries for any one of the 10 counts. This results in the possibility of an extremely large number of decade counter designs. Even if we restrict the design to counters which only progress upward in binary count, the number is still very large.

The weighted counter design, in which each binary has a weight, is a very common design. For example, if we have a 2'421 counter (this is frequently also written as 1242') and the counter is in the count state 1101, the count number can be determined as follows:

Stage	Weight	State	Count
4	$2 \equiv 2'$	1	$1 \times 2 = 2$
3	4	1	$1 \times 2 = 4$
2	2	0	$0 \times 2 = 0$
1	1	1	$\underline{1 \times 1 = 1}$
			7 = total count

In the past, 8421, 2'421, and 4221 decade counters were used, but at the present time the 8421 counter has become the most common. Such standardization has simplified the interface problem to computers, printers, display devices, and similar equipment. These and many other codes are binary coded decimal (BCD). It has become common, however, to represent the 8421 code as BCD. However, to eliminate the confusion, the 8421 code is sometimes represented as NBCD, since it progresses in binary in the normal sequence until 9 (binary 1001) and then returns to 0.

Figure 9-2 is the diagram of an NBCD (BCD) or 8421 decade counter. It makes use of the R–S and J–K properties of the master–slave FF. FF 1 is a toggle FF. FF 2 is a J–K flip-flop with J–K properties determined by the connection from Q of FF 4. FF 3 is a toggle FF and FF 4 is an R–S flip-flop with its R–S properties determined by the states of FF 2 and FF 3. A J–K FF can be used for FF 4 since its R–S properties are the same as those of the master–slave R–S FF.

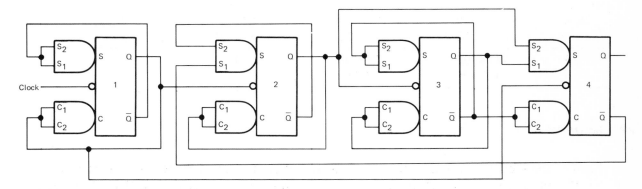

Fig. 9-2. 8421 decade counter.

Assume that the counter is in the 0000 state.

(a) This means that $\overline{Q}$ of FF 4 is at a 1. The 1 connection from $\overline{Q}$ of FF 4 to S_1 of FF 2 puts FF 2 in the toggle mode. FFs 1, 2, and 3 now count in binary up to a decimal count of 7, binary 111.

(b) At this point in the count, the Q outputs of FF 2 and 3 are at a 1. This makes S_1 and S_2 of FF 4 at 1 and the $\overline{Q}$ output of FF 3 at a 0 makes C_2 of FF 4 = 0. At the eighth count FF 1, 2, and 3 all toggle. The Q output of FF 1 is connected to the clock input of FF 4. With S_1 and S_2 of FF 4 = 1 and C_2 of FF 4 equal to 0, FF 4 operates as a clocked R–S flip-flop and FF 4 changes to the 1 state.

(c) The counter is now in the 1000 count state ($\equiv$ decimal 8). The count state is designated by the state of FF 4, FF 3, FF 2, and FF 1 in order. At the ninth count FF 1 toggles and the count state is 1001.

(d) At this count $\overline{Q}$ of FF 4 is at 0. $\overline{Q}$ of FF 4 is connected to S_1 of FF 2, and this 0 input will prevent FF 2 from toggling (J–K properties of FF 2) at the tenth count when FF 1 toggles. The ninth count has made S_1 and S_2 of FF 4 have 0 inputs and the C_1 input of FF 4 has a 1 input after the ninth count. When FF 1 toggles at the tenth count and goes to 0, this clock input to FF 4 makes FF 4 go to a 0 (R–S properties of FF 4). The counter is now back to 0000 after the tenth count.

Mod-N counters have states which are not used. The possibility exists, however, that the counter may get into one of the unused states. Noise, for example, may cause one of the stages to change its state at an undesired time. When the counter is first turned on, each stage arrives at its initial state due to its own internal characteristics. Let us consider what can happen if the counter happens to get into an unwanted state.

One possibility is that it may progress into one of the desired states. It then continues to operate properly. The other possibility is that it continually progresses and recycles through a series of unwanted states in a subloop. Provisions then have to be made in the design of the counter to take it into a wanted state.

EQUIPMENT REQUIRED

CRO, dc and calibrated.

dc power supply, +5 volts at 100 mA.

SWG, positive-going, 5-volt pulses at 10 kHz or 50-μs single pulse.

0.01-μF capacitor, paper or ceramic, 25 volt to 400 volt.

IC type 7472 AND gated J–K FF. Four required.

IC type 7490, decade counter.

IC manufacturers' part numbers.

Type	Motorola	Fairchild	Texas Instruments	National Semiconductor
7472	MC7472P MC7472L	7472PC 7472DC	SN7472N SN7472J	DM7472N
7490	MC7490P MC7490L	7490DC 7490PC	SN7490N SN7490J	DM7490N

PRELIMINARY

Each of the mod-N counters in this experiment will be studied with CRO wave patterns.

Set the pulse generator repetition rate or the CRO horizontal sweep rate so that 10 cycles of the incoming frequency occupy *exactly* 10 horizontal divisions of the CRO. Adjust the horizontal position so that the vertical side of the pulse falls exactly on a vertical CRO line of the graticule. Use negative slope, internal trigger, dc coupling for the trigger, and AUTO mode. DO NOT CHANGE THE REPETITION RATE OF THE SWG OR THE CRO HORIZONTAL SWEEP RATE DURING THE EXPERIMENT.

In the preliminary check of the operation of the counter it is best to use internal trigger.

However, to obtain the correct pattern, the CRO *must be triggered externally from the Q output of the final stage with negative slope.*

The experiment requires that the count states of each of the counters studied in this experiment be determined. The count state always begins with the state of the final stage of the counter and is expressed as a binary number corresponding to the state of the stages, as shown in figure 9-3.

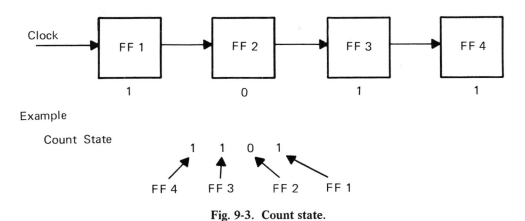

Fig. 9-3. Count state.

EXPERIMENTAL PROCEDURE

1. Mod-3 counter

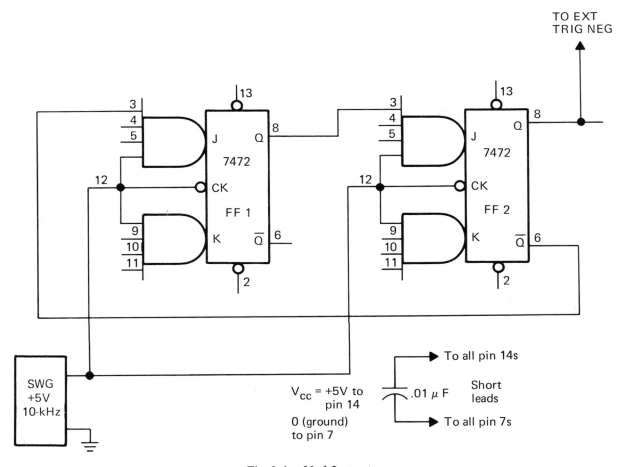

Fig. 9-4a. Mod-3 counter.

Check the Q output of FF 2 to see if the counter divides by 3. Use internal trigger. On the curve sheet of figure 9-4b sketch the wave shapes as they appear at the following points:

Use:

NEG TRIGGER
AUTO
EXT TRIGGER from Q output of final stage
dc TRIGGER

(a) Clock
(b) FF 1 Q output
(c) FF 1 $\overline{Q}$ output
(d) FF 2 Q output
(e) FF 2 $\overline{Q}$ output

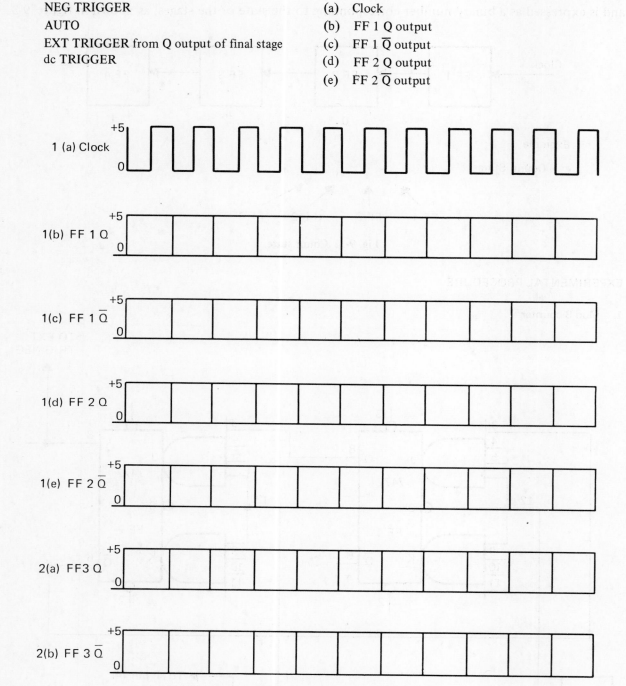

Fig. 9-4b. Wave patterns: Mod-3 and Mod-6 counters.

2. Mod-6 counter

Figure 9-5 is the circuit for this counter. Note that it is exactly the same as that of figure 9-4a except for the addition of a T flip-flop after FF 2 of figure 9-4a.

114

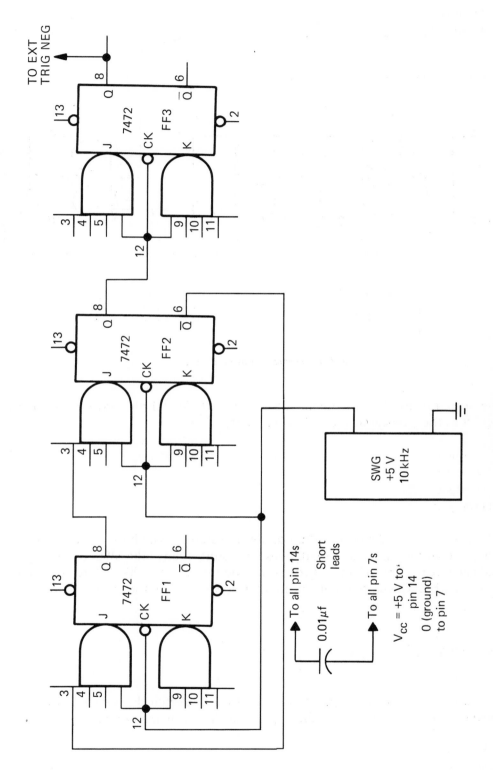

Fig. 9-5. Mod-6 counter.

Check the output of FF 3 for division by 6. On the same curve sheet used for part 1, figure 9-4b, draw the wave patterns as they appear at the following points:

(a)   FF 3 Q output

(b)   FF 3 $\overline{Q}$ output

3.   Decade counter 2'421 (or 1242')

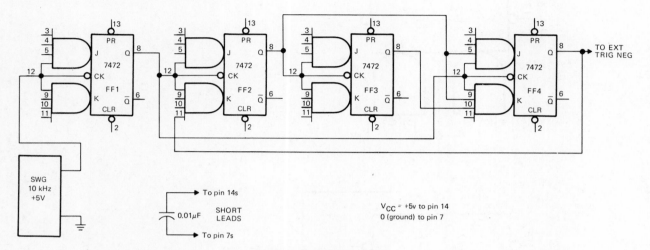

Fig. 9-6. Decade counter 2'421.

Check the output of FF 4 for frequency division by 10. Sketch the wave shapes on the curve sheet of figure 9-8 as they appear at the following points:

(a)   FF 1 Q output

(b)   FF 2 Q output

(c)   FF 3 Q output

(d)   FF 4 Q output

Note: One of the unused states of this counter is a state in which the counter will not count. If, after checking the wiring, the counter still does not count, change the SWG to single pulse, momentarily ground each of the CLR terminals to set the counter into the 0000 state, and then change the SWG back to 10 kHz. Check to see if the counter now counts in the correct manner (output at FF 4 ÷ 10).

4.   Decade counter NBCD 8421.

Check the output of FF 4 for frequency division by 10. Sketch the wave shapes on the curve sheet figure 9-8 on p. 118 as they appear at the following points:

(a)   FF 1 Q output

(b)   FF 2 Q output

(c)   FF 3 Q output

(d)   FF 4 Q output

116

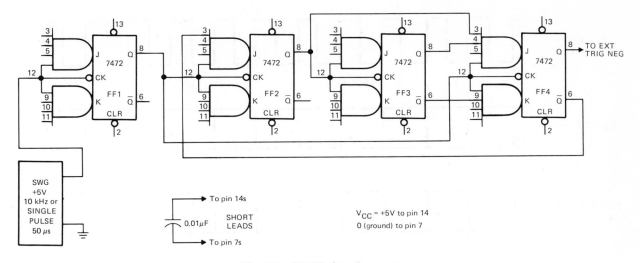

Fig. 9-7. NBCD decade counter.

5. Unused States of the NBCD Counter.

This part of the experiment uses the same experimental set-up as in figure 9-7.

(a) Change the pulse generator to a single pulse of 50 μs.

(b) Reset the counter to the 0000 state using the CLR inputs.

(c) Set the counter to the 1010 state using the appropriate PR and CLR inputs. Note that the state of FF 4 is that corresponding to the first bit in the state of the counter. (1010   FF 4)

(d) Using the CRO (INT TRIGGER) measure the Q outputs of the four FFs. Your results should agree with the operation performed in 5(c) above.

(e) Apply a single pulse, read the state of the counter and tabulate in Table 9-1E. Continue until you either reach a normal counter state based upon the results of the wave patterns of parts 4(a), (b), (c), and (d) or you return to an unused counter state.

Table 9-1E.

Pulse	Count State
0	1010

117

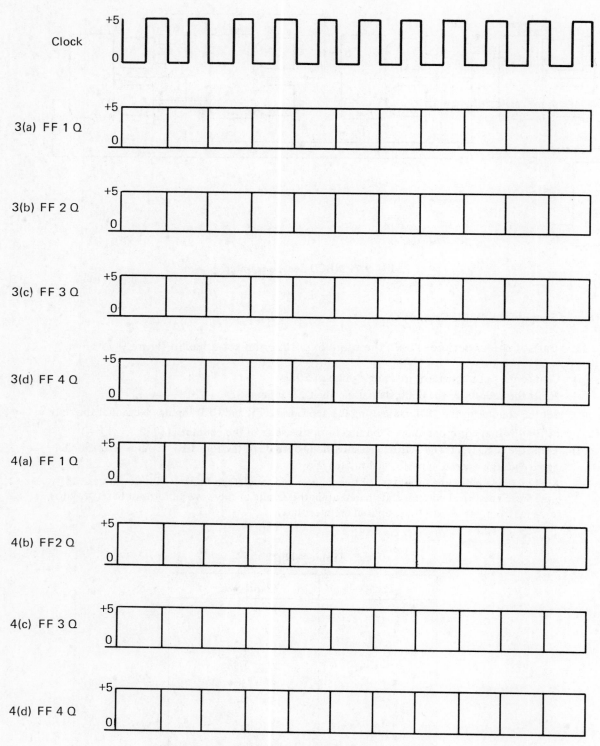

Fig. 9-8. Wave patterns, part 3 (Decade 2'421) and part 4 (Decade NBCD 8421).

118

The TTL type 7490 counter used in this part of the experiment and in part 7 contains 2 sections; a divide-by-2 section between pins 14 (input) and 12 (output) and a MOD-5 section with input to pin 1 and output of pin 11 with intermediate outputs at pins 9 and 8. See logic diagram Appendix D.

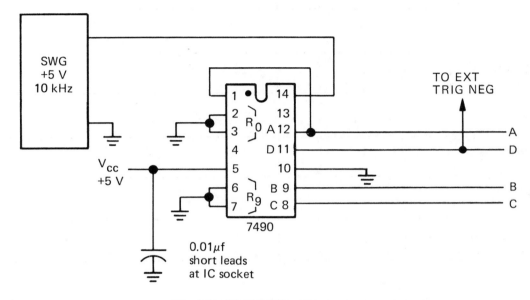

Fig. 9-9. NBCD IC decade counter.

On the curve sheet on figure 9-10 plot the wave patterns that appear at the following outputs:

(a) Q_1 output, A, pin 12

(b) Q_2 output, B, pin 9

(c) Q_3 output, C, pin 8

(d) Q_4 output, D, pin 11

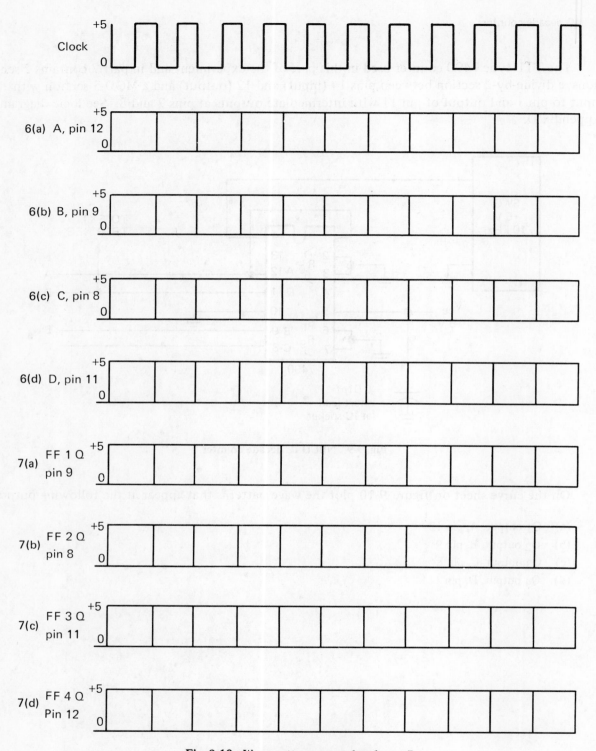

Fig. 9-10. Wave patterns, part 6 and part 7.

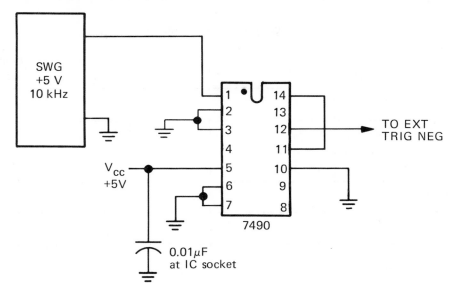

Fig. 9-11. Divide-by-10 counter.

On the same curve sheet as used for part 6, figure 9-10, plot the wave patterns that appear at the following outputs:

(a)   Pin 9

(b)   Pin 8

(c)   Pin 11

(d)   Pin 12

REQUIRED RESULTS

Use logic $1 > 2.5$ volts and logic $0 < 0.5$ volt. The count state is the representation as a binary number of the FF states (Q output) beginning at the left with the state of the final FF.

8.

Complete Table 9-2R for the state of the mod-3 counter of part 1 based upon the wave patterns.

Table 9-2R. Mod-3 counter.

Box	Count	FF 2 Q output	FF 1 Q output	Count State
1	0			
2	1			
3	2			

9.

Complete Table 9-3R for the states of the mod-6 counter of part 2 based on its wave patterns.

Table 9-3R. Mod-6 counter.

Box	Count	FF 3 Q	FF 2 Q	FF 1 Q	Count State
1	0				
2	1				
3	2				
4	3				
5	4				
6	5				

10.

Complete Table 9-4R for the states and the equivalent decimal value of the count based upon the sum of the weighted values for the 2'421 counter of part 3.

Table 9-4R. 2'421 decade counter.

Box	Count	FF 4	FF 3	FF 2	FF 1	Count State	Decimal Count Equivalent
		Weighted Values					
		2	4	2	1		
1	0						
2	1						
3	2						
4	3						
5	4						
6	5						
7	6						
8	7						
9	8						
10	9						

11.

Complete Table 9-5R for the states and the equivalent decimal value of the count based upon the sum of the weighted values for the 8421 BCD counter of part 4.

Table 9-5R. 8421 BCD decade counter.

Count	FF 4	FF 3	FF 2	FF 1	Count State	Decimal Count
	Weighted Values					
	8	4	2	1		
0						
1						
2						
3						
4						
5						
6						
7						
8						
9						

DISCUSSION

1. 10 kHz is the incoming frequency to the counter of part 1. What is the output frequency? How did you determine it? (Hint: Determine the period of each of the waves.)

2. What is the output frequency of the counter of part 2?

3. The BCD 8421 counter of part 4 counts in binary up to the count state of 1001 equivalent to decimal 9. At the tenth count only FFs 4 and 1 change state to return counter to the initial

123

state of 0000. Explain how this takes place. (Hint: Tabulate the J_1, J_2, and J_3, and K_1, K_2, and K_3, inputs after the ninth count. Then, based on the type 7472 truth tables, explain why the next count after 1001 is 0000. (In TTL a not connected input assumes logic 1. See experiment 4.)

4. Based on Table 9-1E, will a momentary transition into an unused state make the counter remain in a stable subloop which is not usable, or will the counter suffer only a temporary malfunction?

5. It is desired to use the 60-Hz power line as the source of a periodic wave whose frequency is 0.01 Hz.
   (a) If you were going to use a single counter, how many stages would be required?

   (b) What is the period of the 0.01-Hz wave?

   (c) You want to use the same decade counter type 7490 used in parts 6 or 7 of the experiment and then build whatever else is needed to obtain the 0.01-Hz frequency by using the 7472 J–K FFs. Show how to obtain the 0.01-Hz frequency. Use a block diagram showing the ICs in each block.

6. Based on the IC interconnection and wave patterns of part 6, what type of counter is the IC decade counter: SW output, 1242′, or NBCD?

OBJECT

    (a)   To study the shift register and its properties.

    (b)   To study ring counters.

    (c)   To study the twisted ring counter.

INTRODUCTORY THEORY

A register is a memory or data store. A *shift register* is a memory in which information is moved or shifted one position at a time in response to a clocked command or shift pulse. The storage element in an IC shift register is the bistable, and a shift register is a cascade combination of bistables. The operation of a shift register is based on the directed R–S properties of R–S flip-flops or the J–K properties of J–K flip-flops (J or K = 1). If the outputs (Q and $\overline{Q}$) of a bistable chain are connected to the J–K inputs (Q to J and Q to K) of the next bistable, a shift pulse will make a following bistable assume or follow the state (data) of the previous stage. Succeeding shift pulses can shift the data to other points in the register.

Circuit arrangements of the shift register permit shifting data in either direction, left or right. Shift registers are therefore called *shift-right* or *shift-left* registers.

Any data bit is shifted one stage for each shift pulse and will appear at a point N stages away delayed by the time of N shift pulses. A shift register can therefore be used as a *delay*.

Serial-type information can be introduced into the first stage of the register 1 bit at a time. After this bit-by-bit information has been transferred into and is stored in the register, the data in all stages can be read out simultaneously in parallel form. This type of shift register is a *serial-to-parallel* data converter. In a similar manner, information can be introduced into all the stages simultaneously in parallel form and read out of the last stage bit by bit by shifting the data toward the last stage. In this form the shift register becomes a *parallel-to-serial* data converter.

When the output of the last stage of a shift register is connected to the input of a shift register, data in the shift register circulates around the register. This is called a *circulating shift register*. It is the electrical equivalent of magnetic drum storage.

A circulating shift register in which *only one* stage is in a state different from all the other stages can be used as a *ring counter*, since the uniqueness of the position of this stage within the ring is determined by and is a measure of the number of shift pulses which are being counted. Such a counter has the advantage of being automatically decoded.

When the output of a shift register is connected to the input in an inverse manner (Q to K and $\overline{Q}$ to J), the counter that results is called a Johnson, Twisted-Ring, Switch-Tail, or Ring-Tail counter.

The counter fills up pulse by pulse until all stages are in the 1 state and then unfills until all stages are in the 0 state. This process repeats itself after a count of two times the number of stages, resulting in a count capability of 2N, where N is the number of stages. To decode this counter requires only 2-input AND gates.

EQUIPMENT REQUIRED

CRO, dc coupled and calibrated.

+5-volt dc power supply at 100 mA.

10-kHz SWG or Single-Pulse 50 μs, +5 volts.

Switch bank–5 switches per bank

IC type 7400 quad 2-input NAND gate.

IC type 7472 AND gated J-K FF.

3 – IC type 7476 dual J-K FFs with PRESET and CLEAR.

IC manufacturers' part numbers.

Type	Motorola	Fairchild Semiconductor	Texas Instruments	National Semiconductor
7400	MC7400P MC7400L	7400PC 7400DC	SN7400N SN7400J	DM7400N
7472	MC7472P MC7472L	7472PC 7472DC	SN7472N SN7472J	DM7472N
7476	MC7476P MC7476L	7476PC 7476DC	SN7476N SN7476J	DM7476N

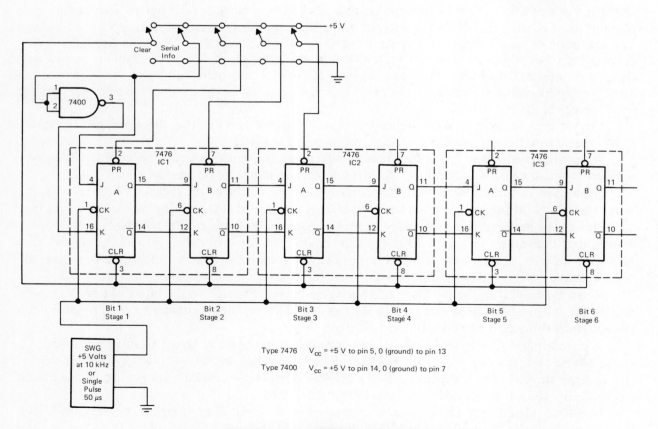

Type 7476 V_{cc} = +5 V to pin 5, 0 (ground) to pin 13

Type 7400 V_{cc} = +5 V to pin 14, 0 (ground) to pin 7

Fig. 10-1. 6 bit shift-right shift register.

All data in this experiment is to be taken *row* by *row*.

1. Shift register

Wire the circuit of figure 10-1 of a 6-bit shift-right register. All switches, the CLR (CLEAR) and the PR (PARALLEL INFORMATION) switches are to be at +5 volts. The serial information switch is to be at 0.

Set the pulse generator to single pulse.

Now, clear all information from the shift register by connecting the CLR switch to ground and then return it to +5.

(a) Parallel in–serial out

Set 1s into both bits of IC 1 by connecting the PR terminals connected to pins 2 and 7 of IC 1 to ground and then returning these PR terminals to +5. Now read the voltages at the Q terminals of the shift register. They should be the values given in Table 10-1E.

Table 10-1E.

IC 1		IC 2		IC 3	
Pin 15	Pin 11	Pin 15	Pin 11	Pin 15	Pin 11
≈ +5	≈ +5	0	0	0	0

Operate the pulse generator which is the shift pulse one pulse at a time. After each pulse, read each of the Q terminal outputs and complete Table 10-2E. Enter data one row at a time.

Table 10-2E.

Shift Pulse	IC 1		IC 2		IC 3	
	Pin 15	Pin 11	Pin 15	Pin 11	Pin 15	Pin 11
0	+5	+5	0	0	0	0
1						
2						
3						
4						
5						
6						

(b) Repeat the previous procedure, clearing the register first, but set 1s into FFA of IC 1 and FFA of IC 2. Complete Table 10-3E.

Table 10-3E.

Shift Pulse	IC 1		IC 2		IC 3	
	Pin 15	Pin 11	Pin 15	Pin 11	Pin 15	Pin 11
0	≈ +5	0	≈ +5	0	0	0
1						
2						
3						
4						
5						

(c) Serial in-parallel out

Clear the shift register. In this part serial information will be set into the shift register by the serial input switch. Set the information switch to the voltage indicated and then apply a *single* shift pulse to the register. Continue by applying the information into the register until all information has been set into the register. Read the voltages and enter in Table 10-4E.

Table 10-4E. Serial in-parallel out.

Serial Information	Shift Pulse Number	IC 1		IC 2		IC 3	
		Pin 15	Pin 11	Pin 15	Pin 11	Pin 15	Pin 11
–	0	0	0	0	0	0	0
+5	1						
0	2						
+5	3						
0	4						
+5	5						
+5	6						

(d)   Circulating shift register

Disconnect the type 7400 from pin 16 and the connection from the serial information switch to pin 4 of IC 1A.

Connect pin 11 of IC 3 to pin 4 of IC 1.

Connect pin 10 of IC 3 to pin 16 of IC 1.

Clear the shift register.

Introduce a 1 into IC 1A and a 1 into IC 1B by means of the appropriate PR inputs.

Change the pulse generator to 10 kHz and view the wave pattern at each of the Q outputs on the CRO and sketch on figure 10-2. The pulse generator pattern should be as shown.

For the CRO use external + trigger from the Q output of the first bit (pin 15 of IC 1).

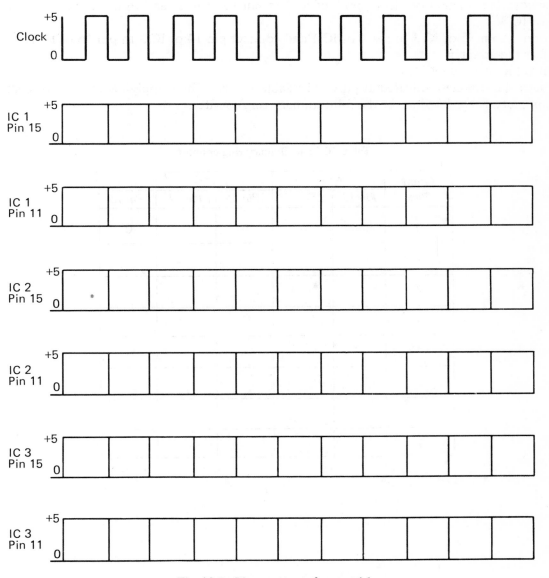

Fig. 10-2. Wave patterns for part 1d.

2. Ring counter

This counter will use the same circuit as in part 1(d), except that it will be modified to have five stages. Whereas in part 1 the pulses applied to the clock inputs of the IC flip-flops were the shift pulses, in this part the pulses applied to the ring counter are the pulses which are counted.

(a) Quinary counter (43210), single-pulse operation

Change the pulse generator to single pulse.

Remove the connections from pin 4 of IC 1 to pin 11 of IC 3, and from pin 16 of IC 1 to pin 10 of IC 3.

Remove the connections from pin 9 of IC 3 to pin 15 of IC 3, and from pin 12 of IC 3 to pin 14 of IC 3.

Connect pin 15 of IC 3 to pin 4 of IC 1 and connect pin 14 of IC 3 to pin 16 of IC 1.

Clear the counter.

Introduce a 1 into IC 1A.

Check the voltages indicated at pulse 0 in Table 10-5Ea. Then, applying a single pulse at a time, complete Table 10-5Ea. for the 5-bit quinary ring counter.

Table 10-5Ea. Quinary ring counter.

Count Pulse	IC 1		IC 2		IC 3
	Pin 15	Pin 11	Pin 15	Pin 11	Pin 15
0	$\approx +5$	0	0	0	0
1					
2					
3					
4					
5					
6					
7					

(b)   Unused states of the ring counter

Clear the counter.
Introduce 1s into IC 1A and IC 2A.
Complete Table 10-5Eb using single pulses from the SWG.

Table 10-5Eb. Quinary ring counter—Unused states.

Count Pulse	IC 1		IC 2		IC 3
	Pin 15	Pin 11	Pin 15	Pin 11	Pin 15
0					
1					
2					
3					
4					
5					
6					
7					

(c)   Wave patterns—Quinary counters

Clear the counter.
Introduce a 1 into IC 1A.
External positive (+) trigger from pin 15 of IC 1.
Change the pulse generator to 10-kHz square wave.
View the wave patterns at the following pin numbers and sketch them on figure 10-3.

(a)   IC 1, pin 15
(b)   IC 1, pin 11
(c)   IC 2, pin 15
(d)   IC 2, pin 11
(e)   IC 3, pin 15

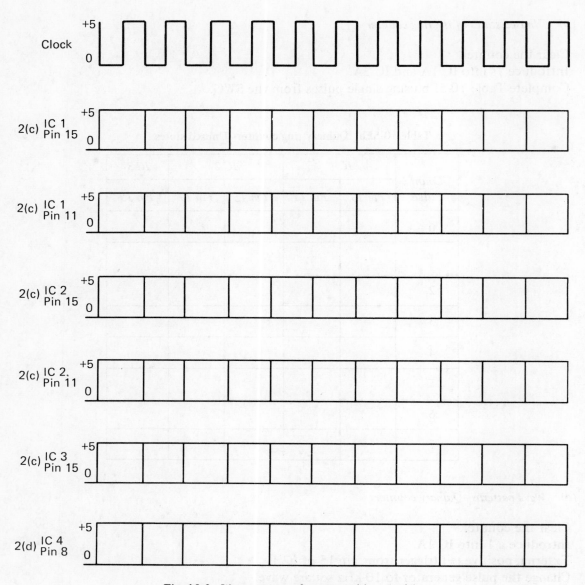

Fig. 10-3. Wave patterns for parts 2(c) and 2(d).

(d)   Biquinary counter (543210 or 5043210)

Add a J–K FF (type 7472) as shown in figure 10-4 to the output of the quinary counter.

Change pulse generator to single pulse.
Clear the counter.
Introduce a 1 into IC 1A.
External negative (–) trigger from the Q output (pin 8) of the FF of figure 10-4 (IC 4).
Change pulse generator to 10-kHz square wave.
View the wave pattern at pin 8 of IC 4. Sketch on the same sheet (Fig. 10-3) used for part (c).

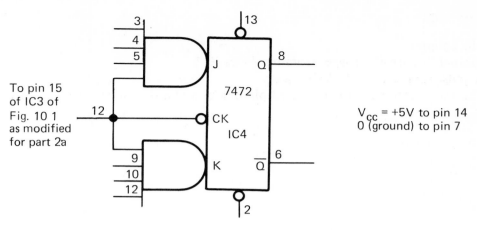

To pin 15
of IC3 of
Fig. 10 1
as modified
for part 2a

V_{cc} = +5V to pin 14
0 (ground) to pin 7

Fig. 10-4. Toggle FF (IC 4) added to binary counter.

3.   The twisted ring or Johnson counter

(a)   Single-pulse operation

Change pulse generator to single pulse.
Disconnect IC 4 from the counter (type 7472).
Change the connections to IC 1 as follows:

Disconnect pin 15 of IC3 from pin 4 of IC1
Disconnect pin 14 of IC3 from pin 16 of IC1
Disconnect pin 15 of IC2 from pin 9 of IC2 and connect this pin 15 to pin 16 of IC1
Disconnect pin 14 of IC2 from pin 12 of IC2 and connect this pin 14 to pin 4 of IC1.

Clear the counter.
Complete Table 10-6E. Apply single pulses.

Table 10-6E. Twisted ring counter.

Count Pulse	IC 1		IC 2
	Pin 15	*Pin 11*	*Pin 15*
0			
1			
2			
3			
4			
5			
6			
7			

(b) Wave patterns

Clear the counter.

Use external (negative) – trigger from pin 15 of IC 1.

Change pulse generator to 10 kHz.

View the wave patterns at the indicated pin numbers and sketch on the curve sheet of figure 10-5.

(a)   pulse generator

(b)   IC 1, pin 15

(c)   IC 1, pin 11

(d)   IC 2, pin 15

(e)   IC 1, pin 14

(f)   IC 1, pin 10

(g)   IC 2, pin 14

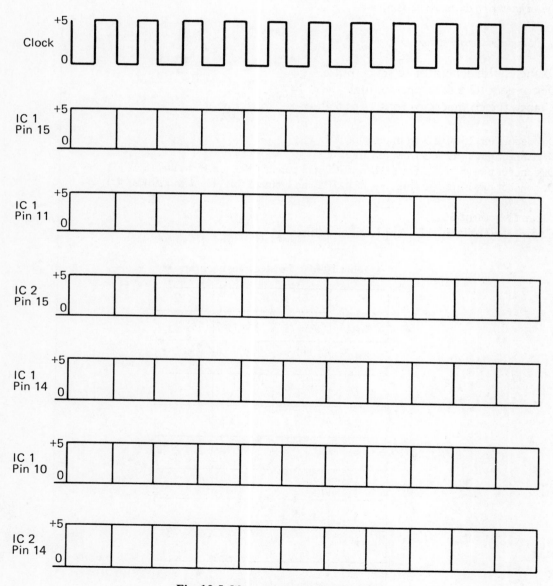

Fig. 10-5. Wave patterns for part 3(b).

Where logic levels are required to complete tables, use 1s and 0s, positive logic $1 \equiv > 2.5$ volts, $0 < 0.5$ volt. For bit number locations see figure 10-1.

1.

(a)   Complete Table 10-1R for the data of Table 10-1E. Use the following equivalents for the bit numbers.

Bit 1   IC 1A, Q output, pin 15
Bit 2   IC 1B, Q output, pin 11
Bit 3   IC 2A, Q output, pin 15
Bit 4   IC 2B, Q output, pin 11
Bit 5   IC 3A, Q output, pin 15
Bit 6   IC 3B, Q output, pin 11

Table 10-1R. Parallel input to 6-bit shift register.

Bit Number	6	5	4	3	2	1
Logic Level						

Complete Table 10-2R for the serial output at the Q output pin 11 of IC 3 (bit 6) based on the data of Table 10-2E.

Table 10-2R. Serial output of 6-bit shift register.

Shift Pulse	0	1	2	3	4	5
Logic Level						

(b)   Complete Table 10-3Ra for the parallel input to the 6-bit shift register of part 1(b) (Table 10-3E, shift pulse 0).

Table 10-3Ra. Parallel input to 6-bit shift register.

Bit Number	6	5	4	3	2	1
Logic Level						

Complete Table 10-3Rb for the serial output at the Q output pin 11 of IC 3 (bit 6) based on the data of Table 10-3E.

Table 10-3Rb. Serial output of 6-bit shift register.

Shift Pulse	0	1	2	3	4	5
Logic Level						

(c)   Complete Table 10-4Ra for the serial input to the shift register based on the data of part 1(c) and Table 10-4E.

Table 10-4Ra. Serial input to 6-bit shift register.

Shift Pulse	1	2	3	4	5	6
Logic Level						

Complete Table 10-4Rb for the parallel output of the shift register based on the data of part 1(c) and Table 10-4E.

Table 10-4Rb. Parallel output for 6-bit shift register.

Bit Number	6	5	4	3	2	1
Logic Level						

5.

(a)   Complete Table 10-5Ra for the count states of the quinary counter (43210 counter) based on the data of Table 10-5Ea. The 43210 counter is a weighted 5-bit coded counter in which the position of the 1 in the count state corresponds to the weight assigned to its position. The position in the code and the location of the stages in figure 10-1 are reversed.

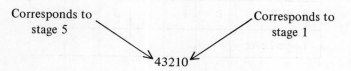

Table 10-5Ra. Count state—quinary counter.

Count Number	Count State
0	00001
1	
2	
3	
4	

(b)   Using the wave patterns of parts 2(c) and 2(d), complete the following state table,

Table 10-5Rc, for the 543210 counter. Also determine the count for each row by adding the weights for each state and show this in the last column.

Table 10-5Rc. State table—543210.

Wave Pattern Box	IC4	IC3	IC2		IC1		Count (Obtain from sum of weighted values)
	Q output	Pin 15	Pin 11	Pin 15	Pin 11	Pin 15	
	Weight						
	5	4	3	2	1	0	
1							
2							
3							
4							
5							
6							
7							
8							
9							
10							

(c)   Give the count state values in Table 10-6R for each of the counts of the Twisted Ring (Johnson) counter of part 3, based on either the values of Table 10-6E or the wave patterns of part 3(b). Use the same state order in going from left to right as that obtained in the data.

IC 1 pin 15   IC 1 pin 11   IC 2 pin 15

↘   ↓   ↙

Bit 1   Bit 2   Bit 3

Table 10-6R. Twisted ring counter—Count states.

Count Number	Count State
0	
1	
2	
3	
4	
5	

DISCUSSION

1. Compare Table 10-1R with Table 10-2R, Table 10-3Ra with Table 10-3Rb, and Table 10-4Ra with Table 10-4Rb, and comment about the ability of a shift register to:
   (a) Shift data.
   (b) Convert binary data from parallel to serial form.
   (c) Convert binary data from serial to parallel form.

2. Based on the wave patterns of part 1(d), by how much is the data in IC 1, pin 15 output delayed until it gets to IC 3, pin 11?

3. A 100-bit shift register has data introduced into it in serial form. The data is being shifted with 50-kHz shift pulses.
   (a) How long will it take the data to reach the 70th bit?

   (b) How long will it take the same data to completely circulate through a 100-bit circulating shift register and appear in serial form again at the output?

4. Based on the data of part 2(b) and Table 10-5Eb, does a forbidden state in a ring counter continue or does it automatically arrive at a correct state? Explain the experimental results to illustrate your answer.

5. Explain the relationship between the count states of the quinary ring counter Tables 10-5Ea and 10-5Ra and the wave patterns of part 2(c). Explain why it is necessary to trigger the CRO as indicated to obtain the correct relationship.

6. A ring counter has developed a fault and the ring has opened. As a result the CRO patterns obtained as in part 2(c) cannot be obtained. Explain how you would determine at which stage the ring was open by the use of a single-pulse generator.

7. What is the relationship between the input frequency and the frequency of the wave seen at any of the Q outputs of the five-stage ring counter of part 2?

8. What is the angular phase relationship between successive outputs of the three-stage twisted ring counter of part 3(b)?

9. What is the relationship between the input frequency and the frequency of the wave that is developed at any of the outputs of the three-stage twisted ring counter of part 3(b)?

10. Which three outputs of the counter of part 3(b) would you select to give a three-phase output?

11. To get a three-phase 60-Hz output of the twisted ring counter of part 3(b), what signal frequency would you put in?

12. Explain how a shift register and parallel adder are used in the arithmetic operation of multiplication.

OBJECT

To study

(a) The transistor astable.
(b) The IC astable.
(c) The IC ring oscillator.

INTRODUCTORY THEORY

THE ASTABLE MULTIVIBRATOR

Figure 11-1 shows the circuit diagram of a transistor multivibrator. As can be seen, two transistors are used in this circuit. Whereas in a transistor bistable the transistors were dc coupled, leading to stable conditions until disturbed by an external signal, the transistors in figure 11-1 are capacitor coupled. This results in two quasistable states, with Q_1 conducting and Q_2 off, and then Q_1 cut off and Q_2 conducting. This interchange of conduction and cutoff condition occurs without any external signal. As will be shown shortly, the time each transistor is in the cutoff condition is determined by the time constant of the resistor and capacitor connected to its base. The circuit acts to automatically charge and discharge the coupling capacitors.

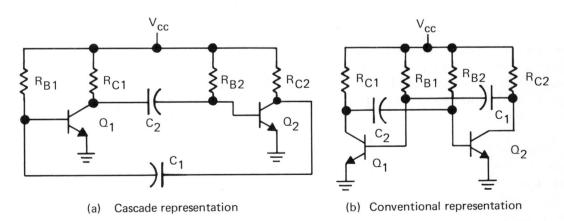

(a) Cascade representation   (b) Conventional representation

Fig. 11-1. Astable multivibrator.

141

Consider a capacitor, resistor, and batteries as shown in figure 11-2. If the switch is closed, the voltage across the capacitor is equal to V volts with point A at a voltage of $-V$ volts. If the switch were open, the A voltage would have to be $+V$ volts. Now, start with the switch in the closed position and open it at time $t = 0$. The voltage at point A must go from $-V$ to $+V$ with time constant $\tau = RC$. We will be interested in the time it takes the voltage at A to go from $-V$ to 0 volts. 0 volts is half way between $-V$ and $+V$. The capacitor has charged to half of its final voltage. This means that we have to determine T in

$$0.5 = 1 - \epsilon^{\frac{-T}{RC}} \tag{11-1}$$

The value of T that satisfies Eq. (11-1) is

$$T = 0.7RC = 0.7\tau \tag{11-2}$$

We now have to show how the circuit of figure 11-1 performs the capacitor charging of figure 11-2.

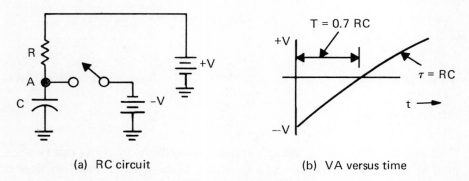

(a) RC circuit   (b) VA versus time

Fig. 11-2. dc charging circuit.

The circuit of figure 11-1 is a saturated circuit. To satisfy the conditions for saturation,

$$\frac{R_{B1}}{R_{C1}} < h_{FE1} \quad \text{and} \quad \frac{R_{B2}}{R_{C2}} < h_{FE2} \tag{11-3}$$

Assume that the V_{BE} of the conducting transistor is equal to 0.

To start, assume Q_1 is conducting and in saturation and Q_2 is cut off, with a voltage across C_2 making the voltage at the base of $Q_2 = -V_{cc}$ volts (this surely cuts off Q_2). Since Q_2 is off, the voltage at the Q_2 collector side of C_1 is equal to V_{cc} and the other side of C_1 connected to the base of Q_1 is at 0 volts. C_1 therefore has a voltage equal to V_{cc} across it.

Now let us consider capacitor C_2. At the Q_1 collector side of C_2 its voltage is equal to 0 and at the other side (at the base of Q_2) its voltage is equal to $-V_{cc}$. We see that it will act like the capacitor C of figure 11-2 at the time its switch was opened.

In figure 11-1, Q_2 is cut off, and therefore its base does not conduct. C_2 charges towards $+V_{cc}$ through R_{B2} as did capacitor C of figure 11-2. When the voltage at the base of Q_2, V_{BE2}, reaches 0, Q_2 is turned on. Both its base and collector conduct heavily. This stops the charging of C_2 towards V_{CC}. The time for the capacitor C_2 to change its voltage from $-V_{CC}$ to $V_{BE2} = 0$ (half way between $-V_{CC}$ and $+V_{CC}$) is given by Eq. (11-4).

$$T = 0.7R_{B2}C_2 \qquad\qquad (11\text{-}4)$$

As Q_2 is turned on, its collector voltage rapidly goes to 0 volts. However, capacitor C_1 was charged with a voltage V_{CC} across it, with the negative side of the voltage at the base of C_1. It takes time for a capacitor to change its voltage (RC time constant). Q_2 turns on too rapidly for the charge on C_1 to change. As a result the voltage across C_1 is maintained and the base of Q_1 is driven to $-V_{CC}$ volts. The operating condition of the two transistors now have been interchanged and the process repeats. Q_1 is now nonconducting for a time given by Eq. (11-5).

$$T = 0.7\ R_{B1}C_1 \qquad\qquad (11\text{-}5)$$

and the total time for a complete cycle is

$$T = 0.7(R_{B1}C_1 + R_{B2}C_2) \qquad\qquad (11\text{-}6)$$

If the circuit has identical components, $R_{B1} = R_{B2} = R_B$ and $C_1 = C_2 = C$,

$$T = 1.4R_BC \qquad\qquad (11\text{-}7)$$

When the off transistor is turned on, its negative-going collector voltage is applied through the coupling capacitor to the base of the on transistor. For a brief period of time both transistors are conducting and in a regenerative mode. This speeds up the transition, which for an astable multivibrator made with small-signal silicon planar transistors takes place in nanoseconds.

Let us now consider the collector wave shape. In figure 11-1 we began with Q_1 on and Q_2 off. Capacitor C_2 had charged to the point where the voltage at its B_2 side was equal to 0. Q_1 had been in saturation and the voltage at the Q_1 collector side of C_2 was essentially 0 (V_{CEsat} of Q_1). At the transition time the voltage across C_2 was therefore equal to 0. After the transition Q_1 is off and the voltage on the B_2 side of C_2 must stay at 0, since Q_2 is on. Since Q_1 is off, C_2 must charge through R_{C1} to V_{CC}. The voltage at the collector side of the off transistor Q_1 rises with a time constant.

$$\tau = R_{C1}C_2 \qquad\qquad (11\text{-}8)$$

When a transistor turns on, R_{sat} is a low resistance and the transition is sharp and very rapid. Figure 11-3 shows the wave shapes of Q_1. The wave shapes of Q_2 are the same except for the timing.

If the collector waveform of figure 11-3 is to be made more square, the time constant $R_{C1}C_2$ has to be reduced. This can only be done by reducing R_{C1}, since changing C_1 would change the time for a cycle. But reducing R_{C1} is in conflict with Eq. (11-3). Nevertheless, by using high h_{FE} transistors, a reasonable approximation to a square wave can be attained. The turn-on edge is always quite sharp and is normally used for timing purposes.

An IC astable multivibrator similar to the transistor astable can be made with two inverter gates by capacity coupling successive gate outputs to inputs. In TTL no input transistor base is brought out to a terminal pin. Therefore, to provide starting gain, it is necessary to

143

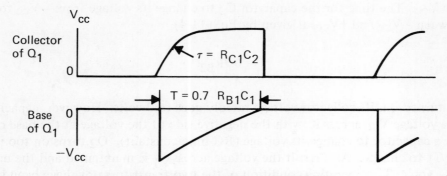

Fig. 11-3. Collector and base waveforms of Q_1.

initially bias the gate transistors in the active region. This can be done with collector feedback resistors as shown in the IC astable of figure 11-4. When operating, the output transistors are, as in the transistor astable, driven into saturation. In the IC astable it is not possible to derive a simple expression for the period.

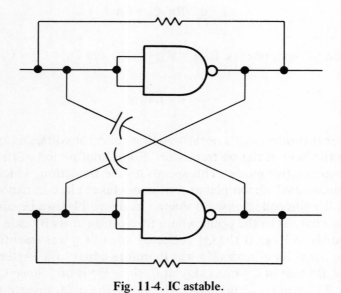

Fig. 11-4. IC astable.

RING OSCILLATOR

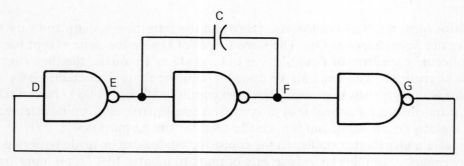

Fig. 11-5. Ring oscillator.

144

Figure 11-5 is the circuit diagram of an IC ring oscillator. Assume that D = 1. Then E = 0, F = 1, and G should be 0. But G and D are connected together. There is a conflict (≡ positive feedback) and hence the circuit is not stable. The levels alternate between 1 and 0 with a period determined by internal gate delays plus a delay determined by the capacitor C.

EQUIPMENT REQUIRED

CRO, dc coupled and calibrated.
Power supply, +5 volts dc at 50mA.
IC type 7400. quad 2-input NAND gate.
IC type 7403. quad 2-input NAND gate, open
   collector.
2 − transistors TIS98.

2 − capacitors 0.01 uF.
2 − capacitors 0.033 uF.
2 − 820 Ω 10% composition resistors.
2 − 1000 Ω 10% composition resistors.
3 − 5600 Ω 10% composition resistors.
2 − 27,000 Ω 10% composition resistors.

IC manufacturers' part numbers.

Type	Motorola	Fairchild	Texas Instruments	National Semiconductor
7400	MC7400P MC7400L	7400PC 7400DC	SN7400N SN7400J	DM7400N
7403	MC7403P MC7403L	7403PC 7403DC	SN7403N SN7403J	DM7403N

EXPERIMENTAL PROCEDURE

For all ICs in this experiment: V_{cc} = +5 volts to pin 14, 0 (ground) to pin 7.

1. Transistor astable

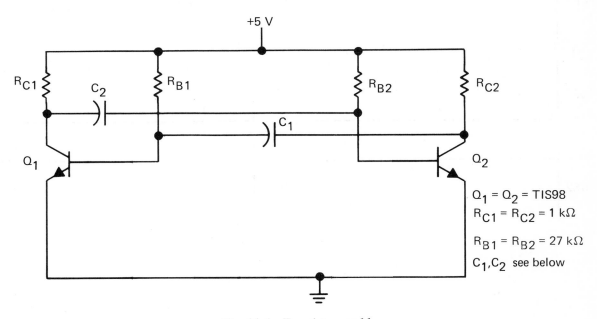

$Q_1 = Q_2 = $ TIS98
$R_{C1} = R_{C2} = 1$ kΩ
$R_{B1} = R_{B2} = 27$ kΩ
C_1, C_2 see below

Fig. 11-6a. Transistor astable.

145

(a) $C_1 = 0.01 \, \mu F$. $C_2 = 0.01 \, \mu F$.

On figure 11-6b sketch the wave forms of V_{CE1}, V_{BE1}, B_{CE2}, and V_{BE2}. Show the voltage amplitudes and the time scale so as to be able to measure the frequency. Use *external TRIG* from the collector of Q_2. At least two complete cycles should be shown.

From the CRO waveforms enter data in the column marked 1a in Table 11-1E. *Do not* proceed until your data has been checked by the instructor.

Table 11-1E.

	1a	1b
Time Q_1 is in saturation		
Time Q_2 is in saturation		

(b) $C_1 = 0.01 \, \mu F$, $C_2 = 0.033 \, \mu F$.

Sketch the wave shapes of V_{CE1}, V_{BE1}, V_{CE2}, and V_{BE2} on figure 11-6b. Particularly note the effect changing C_1 from 0.01 μF to 0.033 μF has on the positive-going wave shape of the collector of Q_2 and indicate this on your sketch. Enter the appropriate data in column 1b of Table 11-1E.

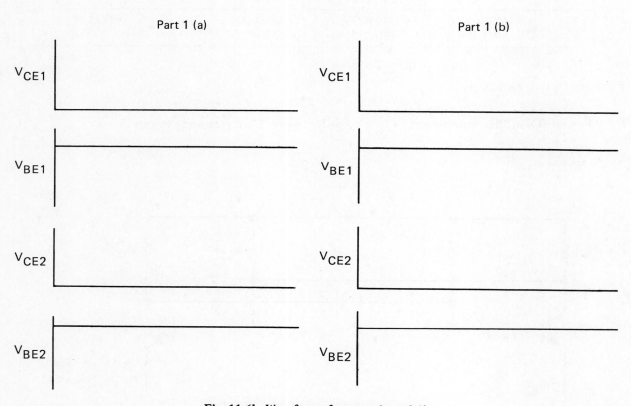

Fig. 11-6b. Waveforms for parts 1a and 1b.

146

2. IC astable

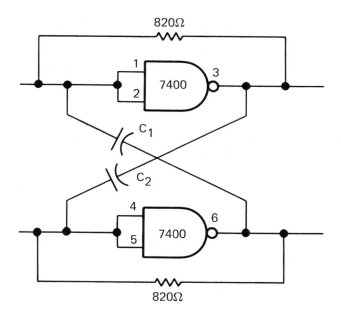

Fig. 11-7a. IC astable.

(a) $C_1 = C_2 = 0.01 \ \mu F$

Sketch the wave shapes at pins 3 and 1 on Figure 11-7b. Use external triggering from pin 3.

Time per cycle = _____

V_3
part 2(a)

V_1
part 2(a)

Fig. 11-7b. Waveforms for part 2.

(b) $C_1 = C_2 = 0.033 \ \mu F$

Time per cycle = _____

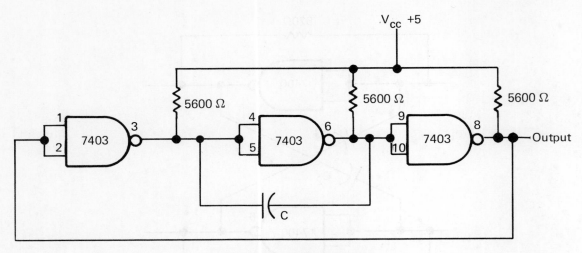

Fig. 11-8a. Ring oscillator.

(a)   C = 0.01 μF. Time per cycle = _____

(b)   C = 0.033 μF. Time per cycle = _____

Sketch the output wave shape for part (b) on figure 11-8b.

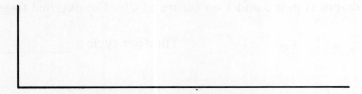

Fig. 11-8b. Waveform for part 3b.

REQUIRED RESULTS

4. From Table 11-1E, complete the following in Table 11-1R.

Table 11-1R.

	Total Time per Cycle	Frequency	Pulse Repetition Rate
1a			
1b			

5. For part 3(a), pulse repetition rate PRR = _____

For part 3(b), pulse repetition rate PRR = _____

6. (a) In part 2, an approximate formula for the total period of the IC astable can be given as

$$P = AC$$

where P is in **microseconds**, C in **picofarads**. and A is a constant. Determine the value of A for the experimental values of part 2(a).

(b) Using the values of C = 0.033 μF in part 2(b), compute the period of oscillation using the value of A from part 6(a) and compare with your experimentally determined results.

7. In part 3 the period of the ring oscillator can be expressed by the formula P = BC (as in part 6). P is in **microseconds** and C in **picofarads**. Determine B for part 3(a) and also for part 3(b) (two determinations for B). If the two values are different, average them to obtain the best value.

DISCUSSION

1. In part 1, explain why the collector negative-going transition is sharp and the positive-going transition is not.

2. In part 1(a), compare the total period of the transistor astable with the theoretical value of 1.4 $R_B C$.

3. In part 1(c), compare the total period with the theoretical value.

OBJECT

To study

(a)   The pulse stretcher/monostable.
(b)   The Schmitt trigger.

INTRODUCTORY THEORY

PULSE STRETCHER

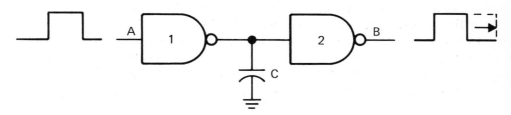

Fig. 12-1. IC pulse stretcher.

The pulse stretcher shown in figure 12-1 is used to extend a pulse width to provide additional time for logic operations. The pulse width at B is longer than that at A because it takes capacitor C time to charge up to the threshold voltage of gate 2.

SCHMITT TRIGGER

The Schmitt trigger, like the bistable, astable, and monostable, is a two-gate regenerative circuit with feedback between input and output. An IC Schmitt trigger is shown in figure 12-2. It is a voltage level detector. When the input voltage is below a voltage level called the upper trip level, UTL, the output voltage is at the 0 logic level, and when it is above the UTL the output voltage is at the 1 logic level. Starting with the input voltage at a high level the circuit will switch to 0 level output at an input voltage called the lower trip level, LTL. The LTL is lower than the UTL and the difference in voltage is commonly referred to as hysteresis, measured in volts:

The circuit is regenerative, and as soon as the trip levels are reached the transition occurs with great speed. Besides its use as a level detector, it can be used to convert a slowly changing signal to one with rapid transitions. This has great utility in pulse-coupled circuitry where the coupling capacitors can be made small. In figure 12-2 resistor R_2 can be used to set the trip points.

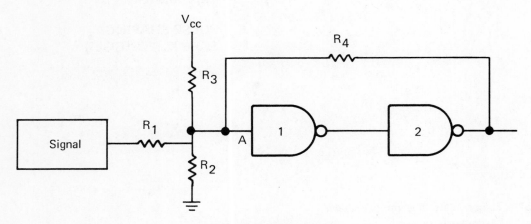

Fig. 12-2. IC Schmitt trigger.

The circuit of figure 12-2 operates in the following manner. When the signal voltage lowers the voltage at point A, the input to gate 1 is below the threshold voltage. This is helped by the 0 level of gate 2 through R_4. When the input signal is at a high enough voltage (UTL) to make the voltage at A equal to the threshold voltage of the gate, gate 1 changes its output level and this in turn makes the output of gate 2 go to the 1 level. This now helps keep point A above the transition voltage (the action is regenerative and takes place rapidly).

Now, if we start to reduce the input voltage, we have to reduce it so that the voltage at point A arrives at the threshold voltage once more. It must be less than the UTL because R_4 is connected now to a 1 level and tends to keep the voltage at A at a higher level. At this voltage, the LTL, the circuit regeneratively switches again rapidly to 0 output level.

EQUIPMENT REQUIRED

CRO, dc coupled and calibrated.

Power supply, +5 volts dc at 50 mA.

SWG, +5 volts at 10 kHz or 50 μs single pulse.

Sine wave oscillator 10 volts p-p at 1000 Hz.

IC type 7403 quad 2-input NAND gate, open collector.

560 Ω resistor, ±10% composition.

680 Ω resistor, ±10% composition.

820 Ω resistor, ±10% composition.

3–1000 Ω resistor, ± 10% composition.

3–5600 Ω resistor, ± 10% composition.

27k Ω resistor, ±10% composition.

500 Ω 3-turn potentiometer.

0.01 μF capacitor.

0.022 μF capacitor.

0.033 μF capacitor.

Silicon small-signal diode, 1N914 or equivalent.

Type	Motorola	Fairchild	Texas Instruments	National Semiconductor
7403	MC7403P MC7403L	7403PC 7403DC	SN7403N SN7403J	DM7403N

EXPERIMENTAL PROCEDURE

For all ICs in this experiment V_{cc} = +5 volts to pin 14 0 (ground) to pin 7.

1. Pulse stretchers

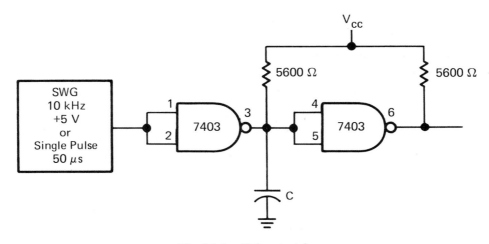

Fig. 12-3a. Pulse stretcher.

Using external + trigger from the SWG, sketch the wave shapes of the SWG and the output at pin 6 for C = 0.033 μF on figure 12-3b. Use external positive sync from the SWG. Plot two cycles of the wave shapes. Complete Table 12-1E.

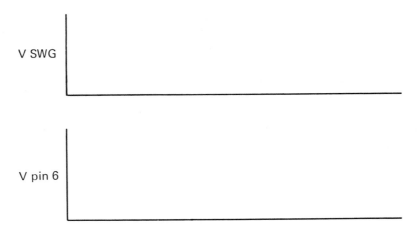

Fig. 12-3b. Pulse stretcher wave shape.

153

Table 12-1E.

C	Pulse Width at Pin 6 (positive section)
0 μF	
0.01 μF	
0.022 μF	
0.033 μF	

(b)   The circuit of part 1(a) was operated with a repetitive 50-μs positive-going pulse generator, but in this part it will be operated on a single-pulse basis. To view the pattern on a single-pulse basis requires that the CRO horizontal sweep controls be set so that a single pulse will trigger it. This is done by means of the level control.

Start by setting the pulse generator at 10 kHz and C at 0.01 μF. View the output at pin 6 using + internal trigger and AUTO mode.

Change the pulse generator to single pulse.

Adjust the CRO level control(s) until a positive-going output pulse is seen for each single pulse from the pulse generator. Complete Table 12-2E.

Table 12-2E.

C	Pulse Width
0.01 μF	
0.022 μF	
0.033 μF	
0.055 μF	
0.065 μF	

2.   Schmitt trigger

For each value of R_A, start with the 500-Ω potentiometer at ground potential. Monitor the output voltage. Slowly increase the input voltage determined by the setting of the 500-Ω potentiometer until the output voltage suddenly changes in amplitude. Call this input voltage the UTP (upper trip point). Increase the input voltage to a value of approximately 0.5 volt more positive than the UTP and then slowly decrease the input voltage until the output voltage suddenly changes. Call this input voltage the LTP (lower trip point). Complete Table 12-3E.

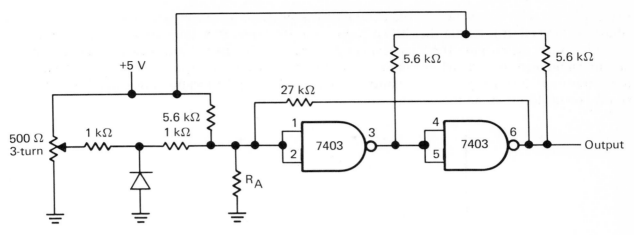

Fig. 12-4. Schmitt trigger.

Table 12-3E. Schmitt trigger.

R_A	UTL	LTP	Hysteresis, volts
1000 Ω			
820 Ω			
680 Ω			
620 Ω			
560 Ω			

Make the following measurements with R_A = 620 Ω.

Input voltage = 0   V at pin 6 = _____

Input voltage = +5   V at pin 6 = _____

3.   "Square" wave from sine wave

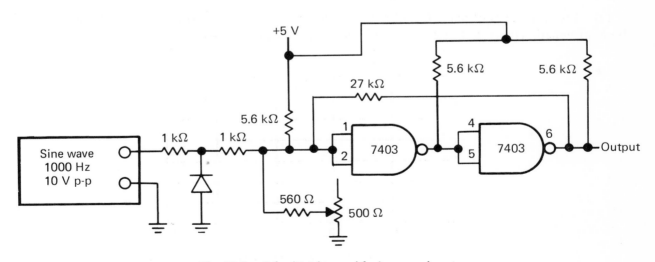

Fig. 12-5a. Schmitt trigger with sine wave input.

(a)   Adjust the 500-Ω potentiometer to give maximum resistance.

(b)   View the input wave pattern. Adjust the sweep speed or oscillator frequency to display

155

two cycles of the input sine wave. Use external + trigger from the sine wave and, if necessary, adjust the level control so that the CRO is triggered at the center (0 voltage level) of the sine wave.

(c) View the output. If a dual-beam or dual-trace CRO is being used, view both input and output simultaneously. Sketch the input and output wave shapes on figure 12-5b. The output is to be drawn on the same plot as the input and in its correct angular location. Be careful to indicate all the voltage locations on the sketch.

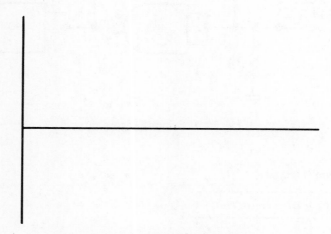

Fig. 12-5b. IC Schmitt trigger waveforms.

(d) Vary the 500-Ω potentiometer and note the effect upon the output voltage. Record your observations in the following space.

REQUIRED RESULTS

4. (a) From the data of part 1(a), plot a curve of the amount the **positive** section of the original pulse is **stretched**. (Increased in width compared to the width for C = 0.)

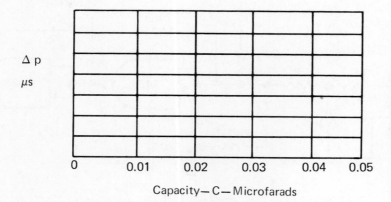

Capacity— C —Microfarads

Fig. 12-6. Stretched pulse vs. capacity.

(b)   From the plot determine the value of the constant N in

$$\Delta P = NC$$

where ΔP is in microseconds and C is in picofarads.

5.   (a)   Plot the data of part 2 on the following graph.

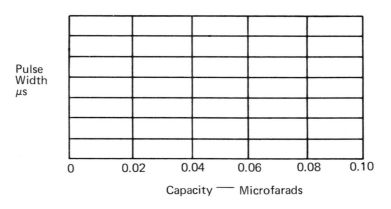

Fig. 12-7. Monostable pulse width.

(b)   From the plot determine the constant H in the formula

$$PW = HC$$

where PW is the stretcher pulse width in **microseconds** and C is in **picofarads**.

6.   Plot the data of part 2 on figure 12-8. Plot one curve for the UTP and another curve for the LTP.

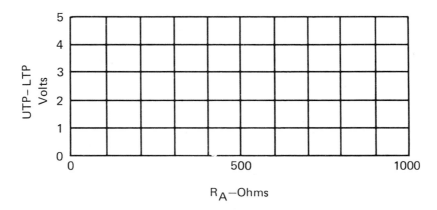

Fig. 12-8. Schmitt trigger, UTP and LTP.

157

DISCUSSION

1. What are some possible applications for the pulse stretcher?

2. Explain the output waveform in part 3(c) and its timing location. Discuss also with reference to the data of part 2.

3. Explain the results of part 3(d).

OBJECT

To study the operations of

(a)   Decoding of electronic counters.

(b)   Encoding and code conversion.

INTRODUCTORY THEORY

1.   Decoding

Electronic counters count in binary or some form of modified binary. Many applications such as display device activation, multiplexing, and timing require conversion from the binary system and its multibit count state to some form of decimal or single-output representation. This conversion process from binary to decimal is known as decoding. This conversion makes use of an AND gate whose inputs are the count state of the counter. Since any divide-by-N counter has N count states, each count can be decoded by such an AND gate and NAND gates are required to decode all of the counter states. For example, the two-stage binary counter shown in figure 13-1 divides by 4

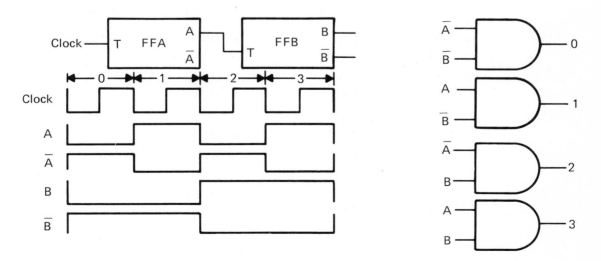

Fig. 13-1. 2-stage binary counter waveforms and decoding.

and has four count states. To decode all four states requires four two-input AND gates with the decoding shown.

When the full count capability of a k-stage binary counter ($=2^k$) is used, decoding all of the count states requires 2^k gates with K inputs to each gate. When the count capability of a k-stage counter is reduced to a value N, N gates (one for each count state) are required to decode the N count states. However, there is some redundancy in the decoding and not all the decoding gates require K inputs. This, therefore, results in decoding simplification.

2. Encoding—Code conversion

Encoding is the opposite of decoding. A binary code is generated for each given decimal value. Encoding requires an OR operation for each bit of the output code. The inputs to the OR gate are determined by the decimal-to-output conversion requirement. Frequently there is a requirement to convert from one binary code to another. A straightforward procedure frequently employed is to decode the initial binary code and then encode the new code.

EQUIPMENT REQUIRED

CRO, dc coupled and calibrated.

dc power supply, +5 volts at 50 mA.

IC type 7404 hex inverter.

2 – IC type 7420 dual 4-input NAND gate.

IC type 7441/74141 BCD to decimal decoder.

IC type 7476 dual J–K FFs with PRESET and CLEAR.

IC type 7490 decade counter.

2 Switch banks, five switches per bank.

5.6kΩ resistor, ±10% composition.

SWG—10-kHz +5V or single pulse 50μs.

IC manufacturers' part numbers

Type	Motorola	Fairchild	Texas Instruments	National Semiconductor
7404	MC7404P MC7404L	7404PC 7404DC	SN7404N SN7404J	DM7404N
7420	MC7420P MC7420L	7420PC 7420DC	SN7420N SN7420J	DM7420N
7441A	MC7441AP MC7441AL	7441PC 7441DC	SN74141N SN74141J	DM7441AN DM74141N
7476	MC7476P MC7476L	7476PC 7476DC	SN7476H SN7476J	DM7476N
7490	MC7490P MC7490L	7490PC 7490DC	SN7490N SN7490J	DM7490N

1.   Decoding

(a)   Two-stage binary counter

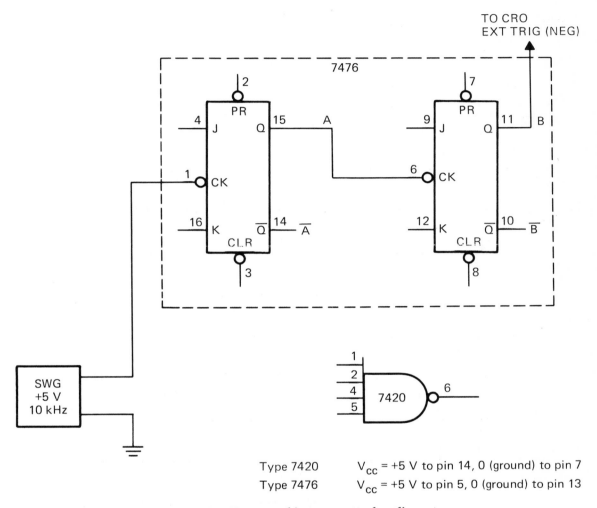

Fig. 13-2a.   Two-stage binary counter decoding gate.

View and sketch on the curve sheet of figure 13-2b the following waveforms:

1.   Output A waveform
2.   Output $\overline{A}$ waveform
3.   Output B waveform
4.   Output $\overline{B}$ waveform

For the following waveforms, pins 4, and 5 of the 7420 NAND gate can be left unconnected.

5.   NAND gate output (pin 6–type 7420) for input $\overline{A}$ to pin 1 (type 7420) and $\overline{B}$ to pin 2 (type 7420) of NAND gate.

6.   NAND gate output (pin 6–type 7420) for input A to pin 1 (type 7420) and $\overline{B}$ to pin 2 (type 7420) of NAND gate.

161

7. NAND gate output (pin 6—type 7420) for input $\overline{A}$ to pin 1 (type 7420) and B to pin 2 (type 7420) of NAND gate.

8. NAND gate output (pin 6—type 7420) for input A to pin 1 (type 7420) and B to pin 2 (type 7420) of NAND gate.

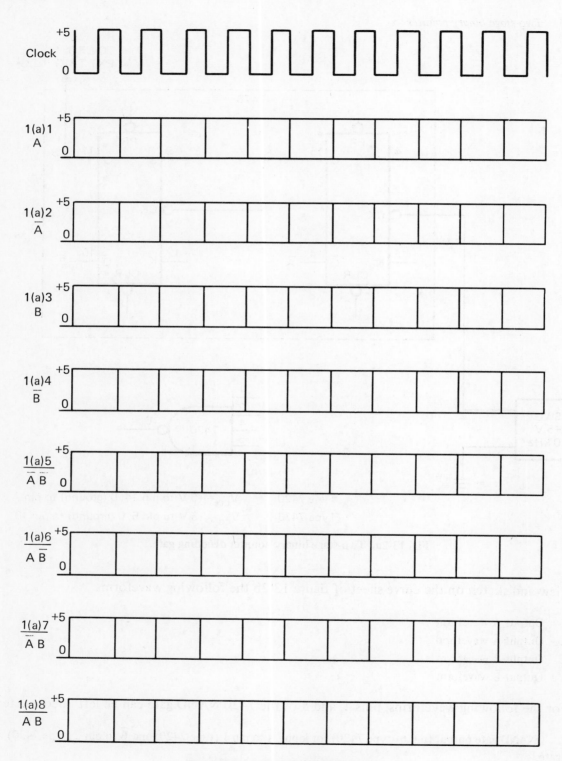

Fig. 13-2b. Waveforms for part 1(a).

(b)   Mod-3 counter

In this part of the experiment, the counter will be decoded for each of the count states, and each decoding operation will be examined for possible redundancy.

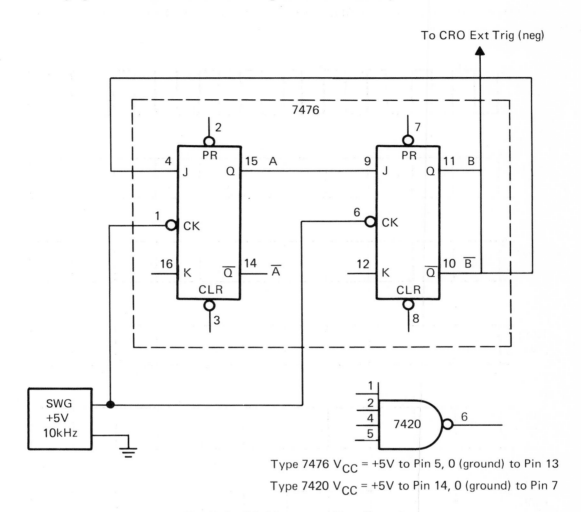

Fig. 13-3a. Mod-3 counter—Decoding gate.

View the sketch on the curve sheet of figure 13-3b the following waveforms:

1.   Output A waveform
2.   Output $\overline{A}$ waveform
3.   Output B waveform
4.   Output $\overline{B}$ waveform

For the following waveforms, pins 4, and 5 of the NAND gate (type 7420) can be left unconnected.

After the decoded waveform has been viewed and sketched, the decoding is to be checked to see if all the inputs are necessary. To do this, remove (one at a time) each of the inputs to the decoding gates and view the pattern. If the decoded pattern changes, the input was necessary and it should be reconnected. These results are to be tabulated in Table 13-1E.

163

5. NAND gate output (pin 6 type 7420) for input $\overline{A}$ to pin 1 (type 7420) and $\overline{B}$ to pin 2 (type 7420) of NAND gate.

6. NAND gate output (pin 6 type 7420) for input A to pin 1 (type 7420) and $\overline{B}$ to pin 2 (type 7420) of NAND gate.

7. NAND gate output (pin 6 type 7420) for input $\overline{A}$ to pin 1 (type 7420) and B to pin 2 (type 7420) of NAND gate.

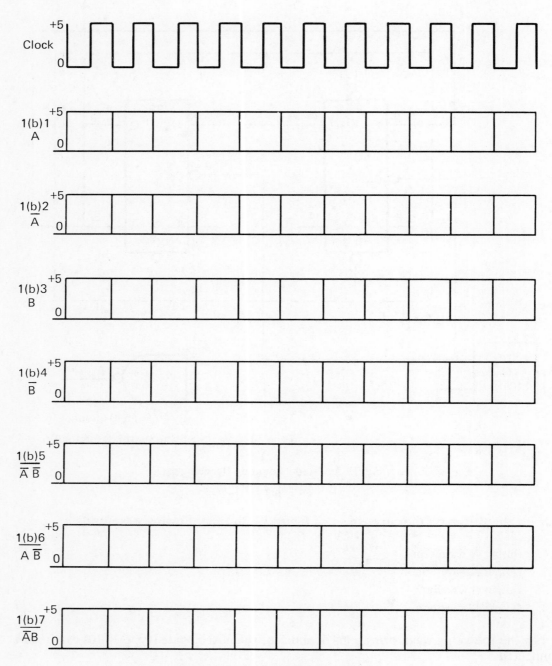

Fig. 13-3b. Waveforms for part 1(b).

Table 13-1E. Mod-3 counter decoding.

Count	Count State Basic NAND Inputs	Necessary Inputs
0	$\overline{A}\overline{B}$	
1	$A\overline{B}$	
2	$\overline{A}B$	

(c)   BCD counter with decimal decoder–single pulse

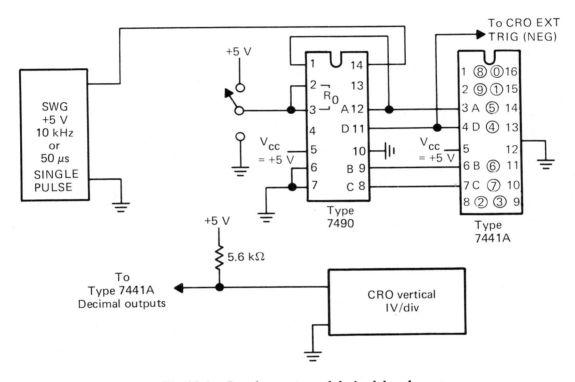

Fig. 13-4a. Decade counter and decimal decoder.

The type 7490 is an NBCD decade counter. The NBCD outputs are D, C, B and A. The type 7441A is on NBCD to decimal decoder. The NBCD inputs are D, C, B and A, and the decimal outputs are encircled. The 7441A outputs have an open collector.

Set the SWG to single pulse. Connect the 5.6-kΩ resistor and CRO input to output 5 (pin 14) of the 7441A decoder. Reset the 7490 counter to zero by connecting both of its pins 2 and 3 to +5 volts momentarily and then returning pins 2 and 3 to ground. Measure the voltage at pin 14 of the 7441A and apply single pulses from the pulse generator. The voltage at output 5 should change at the fifth pulse. Now apply pulses 6, 7, 8, and 9. The outputs for pulses 6, 7, 8, and 9 should be the same as for pulses 0 through 4. If this does not occur, call the instructor.

Repeat in the same manner for output 7 (pin 10). Connect the 5.6-kΩ pull-up resistor to pin 10 of the 7441A.

Change the output of the SWG to 10 kHz. Connect the 5.6-kΩ pull-up resistor and CRO to each of the decimal outputs of the 7441A decoder. View each of the outputs and sketch the output on the curve sheet of figure 13-4b for decimal 0, decimal 5, and decimal 7.

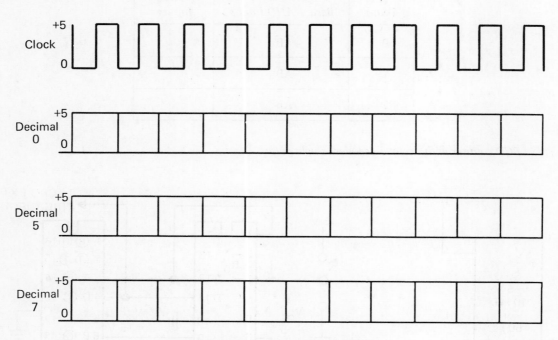

Figure 13-4b. Waveforms for part 1(c).

2.   Encoding—Decimal to excess 3

V_{CC} = +5 volts to pin 14, 0 (ground) to pin 7; all ICs

Table 13-2E.

Decimal	Switch at						D	C	B	A
—	0	1	2	3	4	5	—	—	—	—
0	+5	0	0	0	0	0				
1	0	+5	0	0	0	0				
2	0	0	+5	0	0	0				
3	0	0	0	+5	0	0				
4	0	0	0	0	+5	0				
5	0	0	0	0	0	+5				

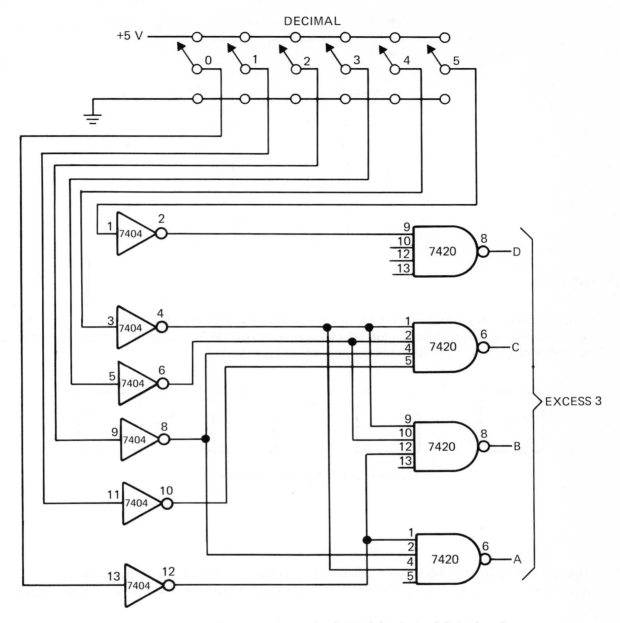

Fig. 13-5. Decimal to excess 3 encoder (DCBA) for decimal digits 0 to 5.

Measure the voltages at outputs D, C, B, and A to complete Table 13-2E.

REQUIRED RESULTS

Based on the results of Table 13-2E, complete Table 13-2R using 1s and 0s for the outputs D, C, B, and A for logic values $1 > 2.5$ volts, $0 < 0.5$ volt.

167

Table 13-2R. Decimal to excess 3.

Decimal	D	C	B	A
0				
1				
2				
3				
4				
5				

DISCUSSION

1. A three-stage binary counter is shown in block diagram in figure 13-6. Draw a logic diagram (decoding gates with required inputs) showing how to decode it at a count of 5 and for a count of 6 (separate gates). The decoded outputs are to be at logic 1.

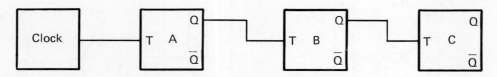

Fig. 13-6. Three-stage binary counter.

2. The counter of figure 13-6 is to be decoded to generate an output of a 1 when the count is at counts of 4 or 5 or both. Show how to do this in the simplest way by drawing a logic diagram.

168

3. Write the excess-3 code for the decimal numbers 0 through 5 and compare with the results of Table 13-2R. Do they agree?

4. Compare Table 13-2R with figure 13-5 and explain how the circuit of figure 13-5 generates the excess-3 code.

5. A two-stage binary counter counts up in normal binary. It is desired to encode it according to the following code. Draw a logic diagram showing how to do this.

Normal Count state	Encoded Output
00	10
01	11
10	01
11	00

6. An "active high" output of a decoder is defined as a decoded output at a high (positive) voltage and an "active low" output is defined as the decoded output at a low positive voltage. Complete the following statements with low or high based upon your experimental data.

Part 1(a) active _____

1(b) active _____

1(c) active _____

OBJECT

(a)   To study the basic principles of semiconductor memories.

(b)   To study the operation of an MSI(TTL) 16-bit memory cell.

INTRODUCTORY THEORY

Semiconductor random-access memories use R–S flip-flops as the building block. By using multigated inputs to control the operation of the FF, a FF in the memory can be randomly selected with information written into and read out of the memory under external control.

The memories used in this experiment enter data into each cell with positive logic, but the information read out appears with negative logic. This is compatible with the entering of parallel data into shift registers. The memories require double rail input data (both levels 1 and 0) and the output is double rail data.

EQUIPMENT REQUIRED

CRO, dc coupled and calibrated.

2 switch banks, five switches per bank.

2 — IC type 7400 quad 2-input NAND gate.

IC type 7403 quad 2 input NAND gate, open
    collector.

IC type SN7481A or MC4005P. 16 bit random access
    memory (RAM).

+5 volt power supply. 50mA regulated.

2 — 1kΩ resistor ± 10% composition.

2 — 5.6k Ω resistor ± 10% composition.

IC manufacturers' part numbers

Type	Motorola	Fairchild	Texas Instruments	National Semiconductor
7400	MC7400P MC7400L	7400PC 7400DC	SN7400N SN7400J	DM7400N
7403	MC7403P MC7403L	7403PC 7403DC	SN7403N SN7403J	DM7403N
7481A MC4005P	MC4005P		SN7481AN SN7481AJ	

1.　2-bit random-access memory

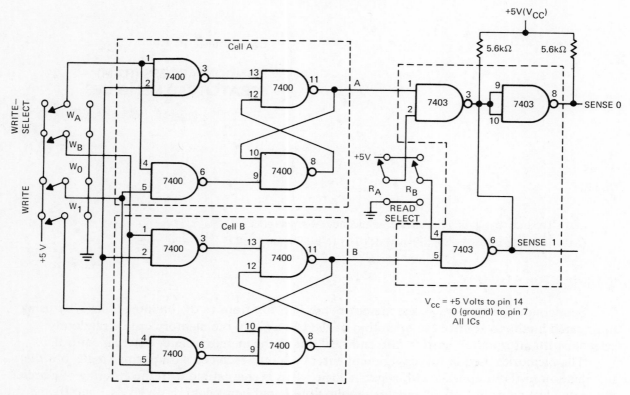

Fig. 14-1. 2-bit random-access memory.

All switches *MUST* be at GROUND initially. To operate the memory, information must first be written into a cell for storage. Information can then be read from the cell.

To write data into a cell, two things must be done:

A.　Select where the information is going, cell A or cell B, by connecting the appropriate select switch (WRITE SELECT A, W_A, or WRITE SELECT B, W_B) to +5 volts.

B.　Write the level of information (0 or 1) into the cell by connecting the appropriate W_0 or W_1 switch to +5 volts. One, but only one, of the WRITE switches must be connected to +5 volts.

CAUTION—ONLY ONE BIT (A or B) and ONLY ONE LEVEL (1 or 0) can be written at a time.

C.　After information is written into a memory cell, return all WRITE SELECT and WRITE LEVEL switches to ground.

D.　A stored 1 makes the "SENSE 1" output less than +0.5 volt and the "SENSE 0" output greater than +2.5 volts. A stored 0 makes the "SENSE 1" output greater than +2.5 volts and the "SENSE 0" output less than 0.5 volt.

E.　To read the stored information, connect the desired READ SELECT (R_A or R_B) switch (only one at a time) to +5 volts. After this is done, read the voltage levels at the SENSE 0 and SENSE 1 outputs. After the stored information has been read, return the READ SELECT switch to ground.

This data is to be taken one line at a time. Store the following information, check the FF output, and sense the stored data by measuring the voltages at the following points to complete Table 14-1E.

Table 14-1E. 2-bit memory.

| | WRITE | | | | READ | | | |
| | Bit A | Bit B | A | B | A | | B | |
					SENSE 1	SENSE 0	SENSE 1	SENSE 0
1	0	0	0	0				
2	0	1	0	1				
3	1	0	1	0				
4	1	1	1	1				

2.   The 16-bit IC RAM

To operate the memory the following procedure should be followed:

A.   All X and Y selection lines and both write inputs are at ground.

B.   To select a bit, connect the appropriate X and Y lines to +5 volts.

C.   To write a 1 in the selected bit, connect the W_1 switch to +5 and return it to 0. This will now make the S_0 output (pin 11) high (greater than +2.5 volts) and the S_1 output (pin 12) low (less than +0.5 volt).

D.   To write a 0 in the selected bit, connect the W_0 switch to +5 and return it to 0. This will now make the S_0 output (pin 11) low (less than +0.5 volt) and the S_1 output (pin 12) high (greater than +2.5 volts).

E.   To read previously stored information, select a bit, as per 2B, and measure the SENSE voltages.

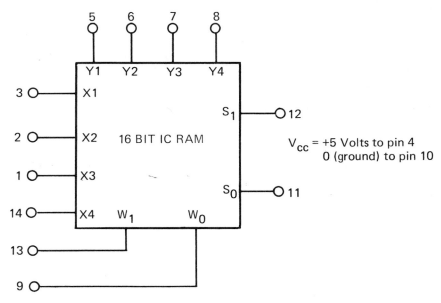

Fig. 14-2. IC RAM

For example, in part 2(a) of the experiment (Fig. 14-3), you will be required to store a 0 in memory cell X3, Y2 (cell 3, 2). Throw the switch connected to pin 1 (X3) to +5 volts, and the switch connected to pin 6 (Y2) to +5 volts. This selects the cell. Then throw the W_0 (WRITE 0) switch to +5 volts and back to ground. This procedure has selected the memory cell 3, 2 and stored a 0 in the cell.

(a) Each box (memory cell) of figure 14-3 has a 1 or a 0 in it. Write this data into each of the cells. As the data is written into each cell, check the SENSE outputs at pins 11 and 12 in accordance with instructions 2C and 2D. **DO NOT turn power off after all data has been written.**

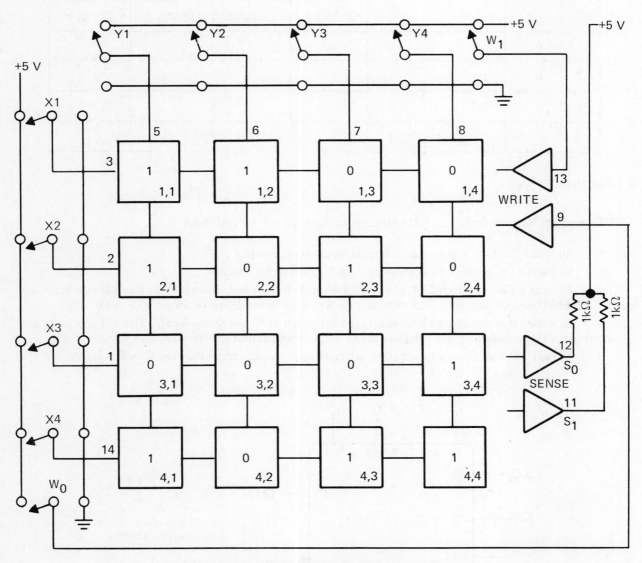

Fig. 14-3. 16-bit RAM—information to be stored.

(b)   Read the information stored in the RAM by measuring the voltages at the S_0 and S_1 outputs and enter the results in the corresponding boxes of figure 14-4. Enter into the boxes the voltages which appear at the S_0 output terminal. **DO NOT turn power off.**

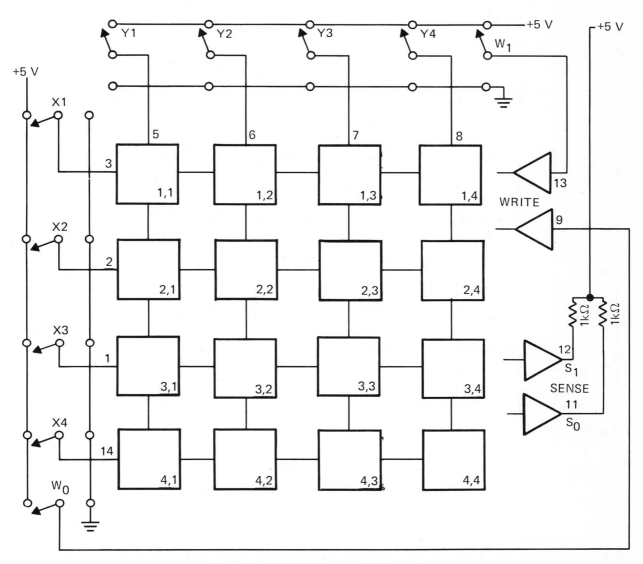

Fig. 14-4. 16-bit RAM--stored information.

(c)   Read the information stored in the RAM again by measuring the voltages at the S_0 and S_1 outputs and tabulate the results in figure 14-5. Enter into the boxes the voltages which appear at the S_0 output terminal. This is to check the memory to see if the memory is DRO (destructive readout) or NDRO (Nondestructive readout).

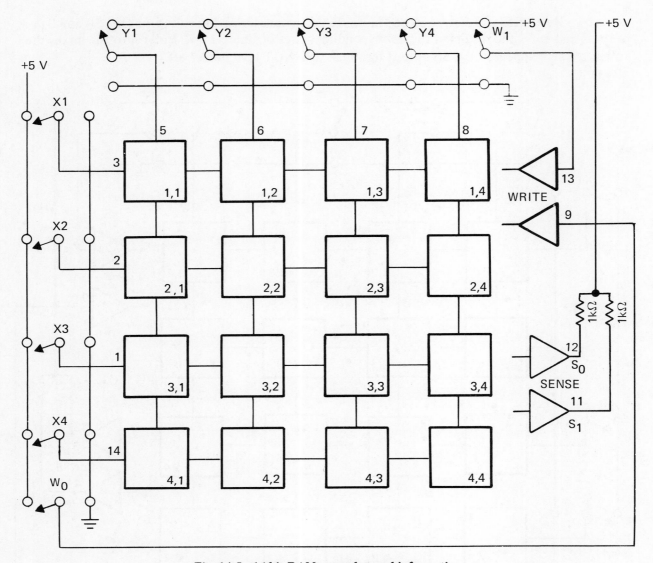

Fig. 14-5. 16-bit RAM—reread stored information.

(d)   Momentarily disconnect the B+ power from the memory. Read the information stored in the RAM and tabulate the results (S_0 output terminal) in figure 14-6. This is to check if the memory is volatile (information lost when power fails) or nonvolatile (information completely retained despite a power failure).

(e)   With all select lines at ground, measure the voltages at the sense outputs.

V at pin 12 _____

V at pin 11 _____

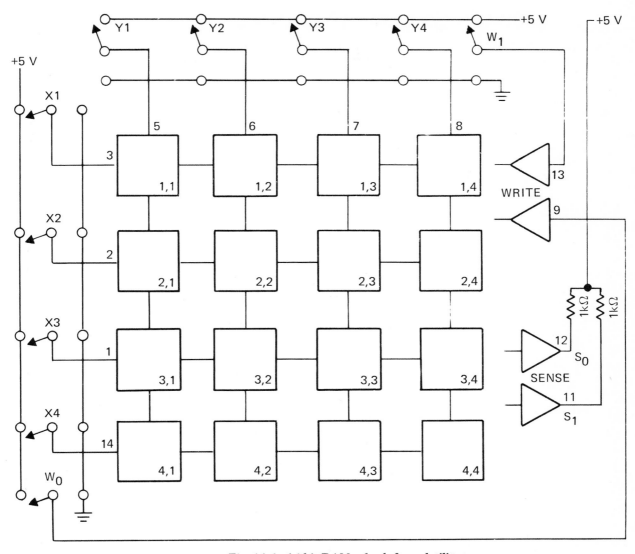

Fig. 14-6. 16-bit RAM—check for volatility.

REQUIRED RESULTS

Use positive logic 1 > +2.5 volts and 0 < +0.5 volt and the sense output logic definition as defined by 1D.

Complete Table 14-1R for the data of Table 14-1E. This table should have 1s and 0s.

Complete Table 14-2R for the data of part 2. The cell location in this tabulation is defined first by the X coordinate and then by the Y coordinate. For example, cell location 2, 3 corresponds to X2 and Y3. This table should have 1s and 0s.

Table 14-1R. 2-bit memory.

	STORED		SENSED	
	Bit A	Bit B	Bit A	Bit B
1	0	0		
2	0	1		
3	1	0		
4	1	1		

Table 14-2R. 16-bit RAM.

Cell Location	Part			
	2(a) Write	2(b) Read	2(c) Read	2(d) Read
1, 1				
1, 2				
1, 3				
1, 4				
2, 1				
2, 2				
2, 3				
2, 4				
3, 1				
3, 2				
3, 3				
3, 4				
4, 1				
4, 2				
4, 3				
4, 4				

DISCUSSION

1. Based on the data of Table 14-2R, is the 16-bit RAM a DRO or NDRO memory?

2. Based on the data of Table 14-2R, is the 16-bit RAM a volatile or nonvolatile memory?

3. Are ferrite core RAMs basically DRO or NDRO?

4. Are ferrite core RAMs volatile or nonvolatile?

5. The Sense 1 output levels of the sense outputs of the RAMs apparently are for negative logic whereas the inputs are for positive logic. Explain why apparent negative logic is needed if the stored data is to be read into a shift register composed of type 7472 J–K FFs. [See also data of part 2(e).]

6. Four IC RAMs are used as four parallel planes as shown in figure 14-1 and provide a storage capability of 16 words of 4-bit capacity with one bit of each word stored in each plane in the same address location. The word $0011 \equiv$ decimal 3 is stored in memory location 1, 1. Explain the steps and show with a simple block diagram how you would use a shift register and the memory to enter a binary word $0110 \equiv 6 \equiv 2 \times 3$ into memory location 1, 2.

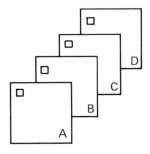

Fig. 14-7. 16-word 4-bit memory—four parallel planes.

7.   The method of figure 14-1 could be extended to include 16 random access cells. After 16 bits of information are stored in this type of memory, how many memory cell READ SELECT switches are required to sense the 16 bits? Compare this with the number of cell select switches required to sense the 16 bits of the 16-bit RAM.

OBJECT

To study the OP-AMP as a

(a)   Precision analog voltage amplifier.
(b)   Multiplying and summing amplifier.
(c)   Comparator.
(d)   Integrator.

INTRODUCTORY THEORY

The operational amplifier, frequently called OP-AMP, is a high voltage gain, high input imped-ance, dc-wide band, inverting amplifier whose output voltage is 0 when the input voltage is 0 volts. The graphic logic symbol for an OP-AMP is shown in figure 15-1.

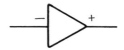

Fig. 15-1. Operational amplifier (OP-AMP).

An OP-AMP is powered with two power supplies (not shown), a positive and a negative voltage supply. ±15 volts is fairly common for the voltages with transistors and ICs. This enables the amplifier to provide signal output voltages up to approximately 75 per cent of the supply voltage and an amplifier with ±15-volt supplies can provide output signal voltages of the order of ±12 volts before the output limits.

Frequently an OP-AMP has two inputs, an inverting and a noninverting input. The logic symbol for this type of amplifier is shown in figure 15-2. In this amplifier the inputs are connected in a differential mode and the output voltage is the amplifier gain multiplied by the voltage differ-ence between the two input signals.

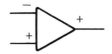

Fig. 15-2. Differential input OP-AMP.

181

Historically, the OP-AMP was first developed for applications in which feedback between output and input was used. When this is done the gain of the amplifier can be made to depend only on the values of the feedback components. With feedback, the amplifier is capable of performing the mathematical operations of addition, multiplication, division, differentiation, and integration upon applied signal wave forms and in this form was the basic element of the analog computer. Logically, only the amplifier with feedback should be called an OP-AMP, but so many other applications have been developed for the amplifier that it is now common practice to call the amplifier alone an OP-AMP.

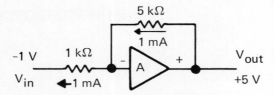

Fig. 15-3. OP-AMP with feedback.

Consider an OP-AMP with feedback as shown in figure 15-3. As will be shown shortly, point A, the input to the amplifier, is at a voltage almost equal to 0, and point A is at a *virtual ground.*

Apply a voltage $V_{in} = -1$ volt. If point A is at ground, 1 mA must flow from A to V_{in} through the 1-kΩ resistor. However, point A has very high impedance and no current can come from the amplifier. The only place current can come from (to maintain current continuity) is from the 5-kΩ resistor. But, since A is at ground, the output voltage must be +5 volts (1 mA $\times$ 5-kΩ).

We can therefore write, neglecting the change in polarity or phase between input and output,

$$\frac{V_{out}}{V_{in}} = \frac{5}{1} = \frac{5 \text{ k}\Omega}{1 \text{ k}\Omega} = 5$$

Hence the gain is determined only by the resistor ratio.

In a practical amplifier the voltage gain might be 50,000 and the input resistance of 500,000 Ω. With 5 volts output the voltage at point A must be

$$\frac{5}{50,000} = 1 \times 10^{-4} \text{ volt}$$

This is close enough to 0 compared to the 1-volt signal and the 5-volt output to justify the assumption that the voltage at A = 0. The input resistance of 500 kΩ is large compared to 1 kΩ and 5 kΩ to justify neglecting its effect (to a first degree of approximation).

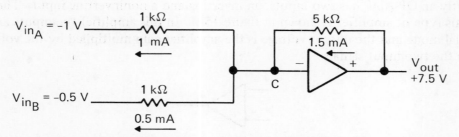

Fig. 15-4. Feedback amplifier with two inputs and equal summing resistors.

Consider figure 15-4, which has two inputs. By extension of the analysis applied to the circuit of figure 15-3, a current of 1 mA must flow into V_{in_A} and a current of 0.5 mA into V_{in_B}.

V_{out} = 7.5 volts = $5(V_{in_A} + V_{in_B})$. Point C acts as a summing point. The circuit of figure 15-4 can be used to add two analog signal voltages.

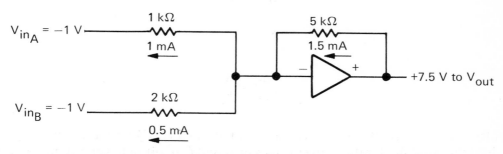

Fig. 15-5. Feedback amplifier with two inputs and unequal summing resistors.

In figure 15-5 equal voltages are applied to summing resistors, but the resistors are not equal. The output voltage is equal to 7.5 volts, obtained as follows:

$$V_{out} = \frac{5 \text{ k}\Omega}{1 \text{ k}\Omega} V_{in_A} + \frac{5 \text{ k}\Omega}{2 \text{ k}\Omega} V_{in_B} = 5(1) + 2.5(1)$$

The analysis of figure 15-5 shows, therefore, that by choice of summing resistors it is possible to change the scale factors by which signal voltages can be multiplied and summed.

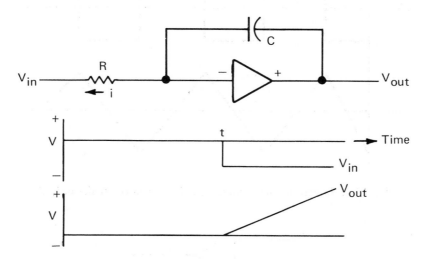

Fig. 15-6. Feedback amplifier with capacitor feedback.

Figure 15-6 shows an OP-AMP with capacitor feedback between the output and the input of the amplifier. Let the input voltage suddenly change to a negative fixed voltage at time t. At this time current must flow from point A to V_{in} with a magnitude

$$I = \frac{V_{in}}{R}$$

As in the previous examples, this current must come from the capacitor C.

But current flowing for a time t means that a charge Q = It must have flown from C. But this could only come about by an increase in output voltage V_{out} given by the expression $Q = C\,V_{out}$. Combining the expression for I, and the values of Q, we obtain

$$Q = \frac{V_{in}}{R}\,t = CV_{out}$$

Hence

$$\frac{V_{out}}{V_{in}} = \frac{t}{RC} \tag{15-1}$$

Equation (15-1) tells us that the output voltage will increase linearly with time, forming a ramp, and continue until the amplifier limits. Larger values of R and C result in lower output voltages and are to be expected. Large values of R reduce the charging rate for C and large capacitors take longer to charge.

The circuit of figure 15-6 is an *integrating circuit*. The voltage at the output depends upon the summation or integral of the charge flowing into V_{in}.

In a square wave of voltage, such as is shown in figure 15-7, is applied to the integrator of figure 15-6, the applied voltage will alternately add charge and subtract charge from the capacitor at a rate based upon Equation (15-1) resulting in a triangular wave output.

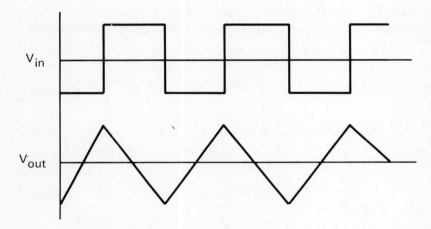

Fig. 15-7. Integrator with square wave input and triangular output.

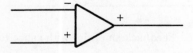

Fig. 15-8. Amplifier with two inputs, inverting and noninverting.

Figure 15-8 shows an amplifier with two inputs, an inverting and a noninverting with no feedback—an open loop configuration. This amplifier has a differential input and amplifies the difference in voltage between the two inputs with high gain. If the minus input is slightly negative with respect to the plus input the output voltage is positive (and probably limiting), and if the minus input is slightly positive with respect to the plus input the output voltage is negative (and probably limiting). The configuration of figure 15-8 can therefore be used as an analog voltage comparator.

If the amplifier limits at 10 volts and has an open loop gain of 50,000, a voltage difference in excess of 10/50,000 = 0.2 mvolt is enough to drive the amplifier into its limiting region. As an analog comparator the amplifier is quite sensitive.

An OP-AMP does not always provide zero output voltage with zero input. Many amplifiers therefore have provisions for a zero output adjustment.

EQUIPMENT REQUIRED

   2 ICs type 741 OP-AMP.
   Resistor, 1000 Ω ±1%, film ¼ watt.
   Resistor, 2000 Ω ±1%, film ¼ watt.
   Resistor, 4990 Ω ±1%, film ¼ watt.
   2 resistors, 470 Ω, ± 10% composition.
   Resistor, 10,000 Ω, ± 10% composition.
   Resistor, 100,000 Ω, ± 10% composition.
   Potentiometer, 500 Ω, 3-turn, 5-watt
   Capacitor, 0.1 uF.
   Capacitor, 0.22 uF.
   Zener diode, 4.7 volts ±10%, 0.5 or 1 watt.
   CRO, dc coupled and calibrated.
   dc power supply, +15 volts, regulated at 50 mA.
   dc power supply, – 15 volts, regulated at 50 mA.
   dc power supply, 0 to 15 volts, regulated at 50 mA. Adjustable.
   SWG, 200 Hz. Max. output impedance 500 Ω at 4 volts p-p.
   Sine wave oscillator. 200 Hz. 30 volts p-p.

IC manufacturers' part numbers.

Type	Fairchild	Motorola	Signetics	National	Texas Instruments
741C	741DC	MC1741CL	N5741A	LM741CD LM741CN-14	SN72741N SN72741J

EXPERIMENTAL PROCEDURE

1.   (a)   The OP-AMP as an analog voltage multiplier

Using the CRO, make voltage measurements at the points indicated to complete Table 15-1E. As each data point is obtained, plot it on the graph of figure 15-10. V_A—horizontal; V_X—vertical.

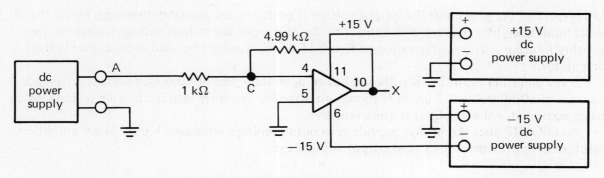

Fig. 15-9a. Voltage multiplier.

Table 15-1E.

V_A	V_C	V_X
CRO 1 volt/div	CRO 0.1 volt/div	CRO 5 volts/div
0	—	
+1	—	
+2	—	
+4	—	
+8	—	
+12	—	
−1	—	
−2	—	
−4	—	
−8	—	
−12	—	

1.   (b)   The OP-AMP as an analog voltage multiplier—CRO X-Y display

Connect the output of a sine wave oscillator whose signal is 30 volts p-p at 200 Hz to the CRO horizontal input and to the 1 kΩ resistor input of the voltage multiplier as shown in figure 15-9b. Center the horizontal display and adjust the horizontal gain so that the pattern displayed occupies six horizontal divisions ≡ 5 volts per division.   View the pattern.   It should agree *exactly* with the plot of figure 15-10. If it does not, call the instructor.

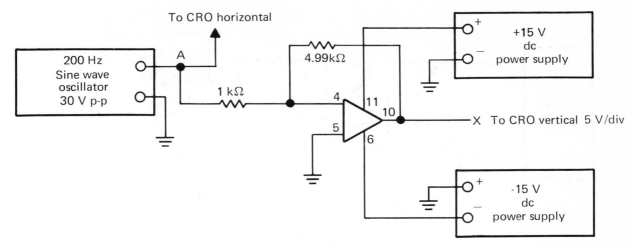

Fig. 15-9b. Voltage multiplier—CRO X-Y display.

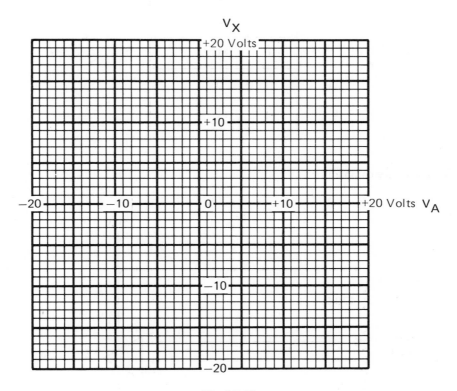

Fig. 15-10.

2.   The OP-AMP as a summer-multiplier

    Measure the voltage at point X with the CRO to complete Table 15-2E. Use CRO sensitivities to give greatest accuracy of readings, but do not use a sensitivity below 0.1 volt/div.

187

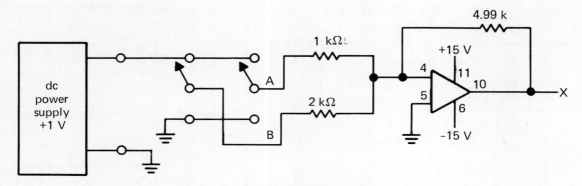

Fig. 15-11. Summer-multiplier.

Table 15-2E.

V_A	V_B	V_X
0	0	
0	+1	
+1	0	
+1	+1	

3. The OP-AMP as a voltage comparator

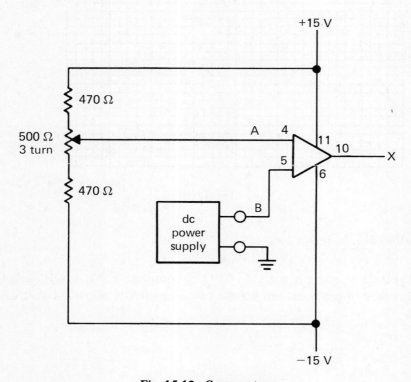

Fig. 15-12. Comparator.

Use the CRO to make all measurements. For each value of V_B in Table 15-3E, vary the setting of the potentiometer to determine the voltage at A which makes the output voltage V_X change rapidly. Measure V_A to within 0.1 volt and complete Table 15-3E.

Table 15-3E.

V_B	V_A
0	
+0.5	
+1.0	
+1.5	
−0.5	
−1.0	

4. Integrator

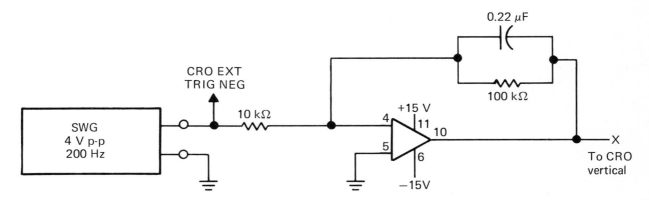

Fig. 15-13a. Integrator.

View the voltage V_X on the CRO and sketch it on the graph of figure 15-13b. The graph should show the voltage amplitudes for V_X and the time scale.

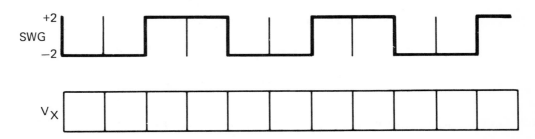

Fig. 15-13b. Integrator waveform.

189

5. Variable pulse-width generator

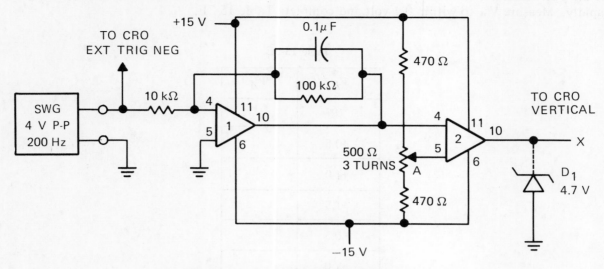

Fig. 15-14a. Variable pulse-width generator.

(a) Zener diode D_1 **NOT** connected to pin 10 of OP-AMP 2.

(1) Adjust the potentiometer to make V_X a square wave. Measure V_A and tabulate on line 3 of Table 15-4E.

(2) Repeat 1 but for the most positive portion to be 25 per cent of a cycle. Enter on line 2 of Table 15-4E.

(3) Repeat 1 but determine V_A for the positive portion to just disappear.

(4) Repeat 1 but for the most positive portion to be 75 per cent of a cycle.

(5) Repeat 1 but for the positive portion to just occupy the complete cycle.

Table 15-4E.

	Positive Per Cent of Cycle	V_A
1	0	
2	25	
3	50 (SW)	
4	75	
5	100	

(b) Sketch the waveform at V_X for the 50 per cent cycle (SW) on figure 15-14b. Show voltages and the time scale.

(c) Connect the 4.7-volt zener diode between point X and ground as shown in figure 15-14a. Vary the potentiometer and note the effect. Sketch the waveforms at V_X for the 50 per cent cycle on the graph below. Show voltages.

190

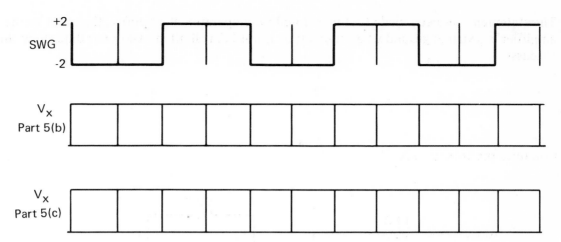

Fig. 15-14b. Waveforms variable pulse-width generator.

DISCUSSION

1. In part 1, what was the measured voltage gain V_X/V_A for

    (a) $V_A = +1$ volts _____   (b) $V_A = +2$ volts _____

    (c) $V_A = -1$ volts _____   (d) $V_A = -2$ volts _____

    How does this compare with the theoretical voltage gain and what determines the theoretical voltage gain and its precision?

2. What were the positive and negative output limiting voltages in part 1? What percentage were these values of the respective power supply voltages?

3. If the open-loop voltage gain of the uA741 is 100,000, what voltage would you expect at point C if $V_X = -10$ volts? Compare this with your measured voltage.

4. The statement has been made that in a closed-loop operational amplifier, the input to the amplifier is a virtual ground when the output is not being limited. Did your data show this? Discuss.

5. Compute the voltage at X.

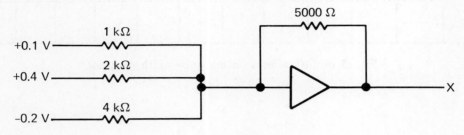

Fig. 15-15.

6. Based on the results of part 3, what general statement can be made about the value of the input voltage at A compared to that at B at which the amplifier changes voltage rapidly?

7. The switch is changed to make the voltage at A equal to +2 volts. After 2.0 ms, what voltage should be expected at V_X? Show all calculations. Compare with the experimental change in integrator voltage of part 4 over a time of 2.0 ms.

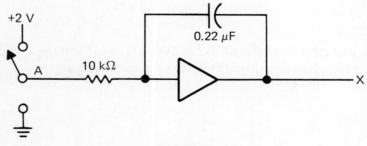

Fig. 15-16.

8.   Explain the operation of the circuit of figure 15-14a with the Zener diode in the circuit.

OBJECT

To study

(a)   The binary weighted ladder method of D/A conversion.
(b)   The comparison method of A/D conversion.

INTRODUCTORY THEORY

Virtually all physical phenomena, voltage, pressure, temperature, velocity, and so on, occur in nature on an analog basis. However, there are distinct advantages to digital representation, such as information storage and data transmission. The requirement exists, therefore, to convert data from an analog basis to a digital representation, A/D conversion, and conversely, from digital to analog, D/A.

In the binary weighted ladder D/A conversion technique, a resistor network is used with values increasing by multiples of 2. These resistors are connected between the summing junction of an OP-AMP and the outputs of the counter stages. This provides currents into the summing junction proportional to the binary weight of the counter stages. The feedback resistor of the OP-AMP can now be selected to provide a scale factor. This will now make the OP-AMP output an analog output voltage equal to the binary weighted sum of the count states of the counter multiplied by the selected scale factor.

The analog output voltage of the D/A converter cannot change continuously, but changes by fixed amounts. The smallest possible change occurs when the 2^0 bit changes. The 2^0 is the least significant bit (LSB). Each higher-order bit changes the analog output voltage by an additional multiple of 2 until the highest-order bit, called the most significant bit (MSB), changes. For example, consider a three-stage binary counter whose maximum count state is $1 \times 2^2 + 1 \times 2^1 + 1 \times 2^0 = 7$. If the scale factor of the OP-AMP is such that the LSB corresponds to a change in 1 volt, the analog output will progress in LSB steps of 1 volt. The MSB = $1 \times 4 = 4$ volts and the analog voltage changes in steps of 1 volt up to a maximum output of 7 volts. To obtain a finer resolution in output voltage, the number of bits must be increased and the value of the LSB decreased.

If the output of a D/A converter is connected to the input of an OP-AMP connected as a comparator and an unknown analog voltage connected to the other input of the OP-AMP, the output voltage of the OP-AMP will be at its limiting value, positive or negative depending on the polarity of the voltage difference between the D/A output and the unknown analog voltage. By determining the counter count state at which the comparator output changes, the count state

195

equivalent to the analog voltage within one bit, the least significant bit (LSB), can be determined. This comparison method, therefore, can be used as an A/D conversion.

EQUIPMENT REQUIRED

CRO, dc coupled and calibrated.

SWG, 1 kHz or single pulse, +5 volts.

Power supply, +15 volts, regulated 50 mA.

Power supply, −15 volts, regulated 50 mA.

Power supply, +5 volts, regulated 50 mA.

IC type 7404 hex inverter.

IC type 7405 hex inverter (open collector).

IC type 7490 decade counter.

2 ICs type 741C OP-AMP

5 Resistors, 1.5 kΩ ±1%, ¼ watt.

Resistor, 2370 Ω ±1%, ¼ watt.

Resistor, 4530 Ω ±1%, ¼ watt.

Resistor, 4750 Ω ±1%, ¼ watt.

Resistor, 10,500 Ω ±1%, ¼ watt.

Resistor, 22,600 Ω ±1%, ¼ watt.

2 resistors, 1 kΩ, ±10% composition.

Resistor, 68 kΩ, ±10% composition.

2 potentiometers, 500-Ω, 3-turn.

Switch bank—5 switches per bank.

IC manufacturers' part numbers.

Type	Motorola	Fairchild	Texas Instruments	Signetics	National Semiconductor
741C	MC1741CL	741DC	SN72741N SN72741J	N5741A	LM741CD LM741CN-14
7404	MC7404P MC7404L	7404PC 7404DC	SN7404N SN7404J	N7404A N7404F	DM7404N
7405	MC7405P MC7405L	7405PC 7405DC	SN7405N SN7405J	N7405A N7405F	DM7405N
7490	MC7490P MC7490L	7490PC 7490DC	SN7490N SN7490J	N7490A N7490F	DM7490N

EXPERIMENTAL PROCEDURE

Power supply connections

Type 7404 V_{CC} = +5 volts to pin 14, 0 (ground) to pin 7

Type 7405 V_{CC} = +5 volts to pin 14, 0 (ground) to pin 7

Type 7490 V_{CC} = +5 volts to pin 5, 0 (ground to pin 10

Type 741C   +15 volts to pin 11, − 15 volts to pin 6

1.   (a)   D/A conversion—Decade BCD

The type 7490 decade counter will count when pins 2 and 3 are connected to ground and can be reset to zero by connecting pins 2 and 3 to +5 volts. Returning pins 2 and 3 to ground will allow the type 7490 to count.

Set the SWG to 1000 Hz and view the output of the type 741C OP-AMP on the CRO. The output should be a nine-step negative-going staircase waveform, with each step having the same height. If you do not obtain the staircase, vary the zero control.

Adjust the horizontal sweep rate so that two staircase waveforms are seen.

Adjust the 500 Ω zero setting control so that the top step of the staircase is at 0 volts.

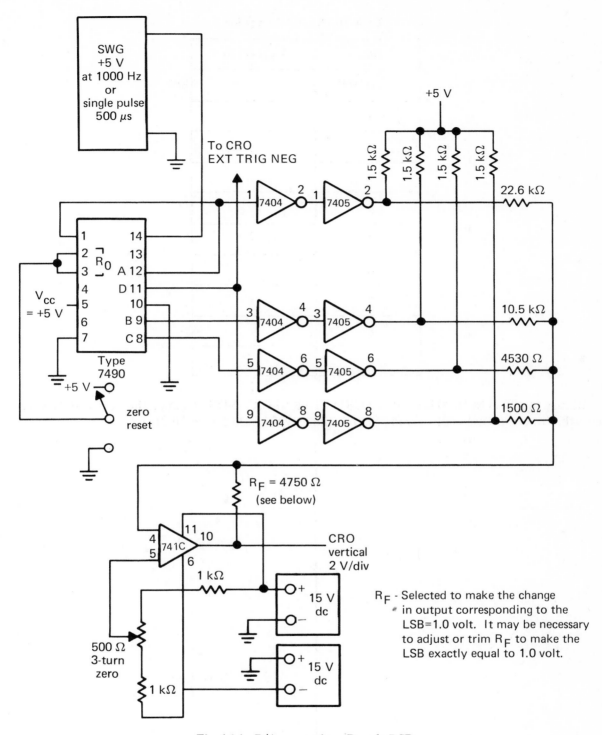

Fig. 16-1. D/A conversion—Decade BCD.

With the resistor values used, the most negative step should give a vertical deflection of −9.0 volts.

Change the SWG to single pulse. Reset the type 7490 decade counter to zero by connecting pins 2 and 3 to +5 volts and then to ground. Change the CRO TRIG to INT.

Measure the output voltage of the D/A converter applying a single pulse at a time and complete Table 16-1E.

197

Table 16-1E. D/A conversion.

Pulse	V_{out}–Volts
0	
1	
2	
3	
4	
5	
6	
7	
8	
9	

Return the SWG to 1000 Hz and the CRO TRIG to EXT NEG. Change the horizontal sweep rate so that one staircase is displayed. Sketch the waveform in figure 16-2(a).

(a)   (b)

Fig. 16-2. Decade staircase waveform.

1. (b) Scale factor change

Change R_F to 2730 Ω. If necessary adjust or trim R_F to make the LSB = 0.5 volt.
Sketch on figure 16-2b and label the pattern and that of part 1(a) with the values of R_F and the scale factors for the LSB.

1. (c) Effect of "opens" in D/A converter

Change R_F to 4750 Ω. (Use value from part 1(a) if changed from 4750 Ω.)
Open the connection between the following pins and note the effect on the staircase pattern. In each case sketch the pattern on figure 16-3.

1. Type 7404 pin 2 and type 7405 pin 1 $\equiv 2^0$. Sketch on figure 16-3a.
2. Type 7404 pin 4 and type 7405 pin 3 $\equiv 2^1$. Sketch on figure 16-3b.
3. Type 7404 pin 6 and type 7405 pin 5 $\equiv 2^2$. Sketch on figure 16-3c.

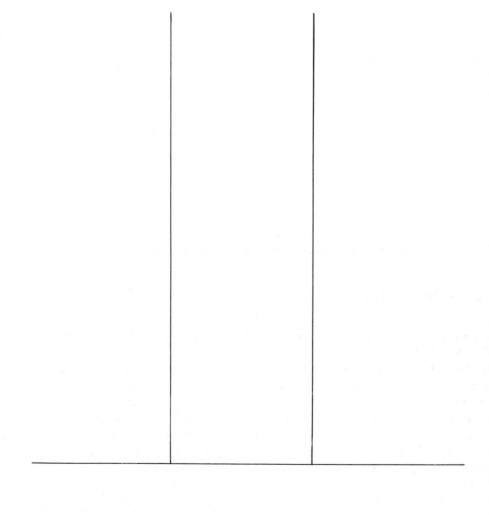

(a)   (b)   (c)

Fig. 16-3. Decade staircase with open errors.

199

1. (d)   Effect of resistor precision error

Shunt the 10.5 kΩ resistor on the ladder with a 68 kΩ ±10% resistor. Sketch on figure 16-4. Remove the 68 kΩ resistor. The sketch should show the pattern with and without the 68 kΩ resistor.

Fig. 16-4. D/A binary ladder—Effect of resistor precision errors.

2. (a)   A/D conversion—Comparison method

This uses the same circuit as part 1 for the D/A conversion with the additional comparator circuitry of figure 16-5.

Change the pulse generator to single pulse. Reset the counter to a count of 0. Set the potentiometer connected to pin 5 (OP-AMP), figure 16-5, to the voltage given in line 1 of Table 16-2E. (Use CRO to measure voltage.) Monitor the output voltage of the OP-AMP of figure 16-5. Apply single pulses to the decade counter until the monitored OP-AMP output changes from one limiting polarity to the other. Measure the counter count state at the type 7405 output pins. (Note: Connecting a lead from the CRO to the type 7490 counter may introduce spurious noise pulses which can be counted by the 7490 counter which is very fast. This is the reason for the measurement at the type 7405).

Repeat this procedure for each of the pin 5 voltages of Table 16-2E, being careful to reset the type 7490 counter to a count of zero in each case.

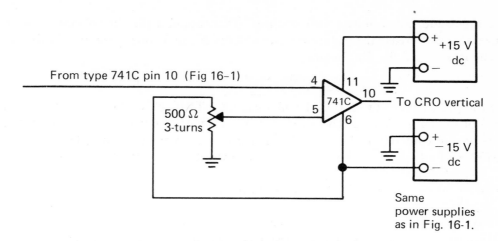

From type 741C pin 10 (Fig 16-1)

500 Ω
3-turns

741C

4   11
        10   To CRO vertical
5
      6

+15 V
dc

− 15 V
dc

Same
power supplies
as in Fig. 16-1.

Fig. 16-5. A/D conversion—Comparison method.

Table 16-2E. A/D conversion.

	Pin 5 Voltage (Fig. 16-5)	Voltage Measurements at Type 7405 Pins after OP-AMP output change Use CRO sensitivity of 1 v/div.			
		Type 7405 Pin Voltages			
		Pin 8	Pin 6	Pin 4	Pin 2
1	−0.5				
2	−1.5				
3	−2.5				
4	−3.5				
5	−4.5				
6	−5.5				
7	−6.5				
8	−7.5				
9	−8.5				

2.   (b)   Resolution limits of the A/D converter

Set the pin 5 potentiometer voltage to − 2.5 volts. Reset the counter to 0. Apply single pulses until the OP-AMP output changes. Now, slowly increase the pin 5 voltage (more negative) until the OP-AMP (Fig. 16-5) output just changes. Read the pin 5 voltage.

V = _____

Apply a single additional pulse to the counter. This should make the OP-AMP (Fig. 16-5) output change. Now, increase the potentiometer voltage at pin 5 (more negative) until the OP-AMP output just changes. Read the pin 5 voltage.

$$V = \underline{\hspace{2cm}}$$

Change the pulse generator to 1000 Hz and use EXT TRIG NEG. (See Fig. 16-1) Check the input to pin 4 of the OP-AMP (Fig. 16-5) to see if it is a 9-volt negative staircase [as in part 1(a)]. Now view the output (pin 10) of the OP-AMP of figure 16-5. Vary the potentiometer connected to pin 5 of the OP-AMP of figure 16-5 and monitor the output of the OP-AMP of figure 16-5.

(1)   What are the output voltage levels of the wave pattern?

$$V_{max} = \underline{\hspace{2cm}} \qquad\qquad V_{min} = \underline{\hspace{2cm}}$$

(2)   Does the rectangular pattern viewed change smoothly or in steps? Sketch the wave pattern for the voltage at pin 5 = –3.5 and –4.5 volts in figure 16-6 (two patterns). Show the time scale. Label on figure 16-6.

Fig. 16-6. Comparator output.

REQUIRED RESULTS

Using positive logic, complete Table 16-2R corresponding to Table 16-2E for the NBCD equivalents of the voltages applied to pin 5. (Fig. 16-5.) The table should have 1s and 0s.

Table 16-2R. NBCD equivalents.

Pin 5 Voltage	Type 7405 Logic Levels			
	Pin 8	Pin 6	Pin 4	Pin 2
−0.5				
−1.5				
−2.5				
−3.5				
−4.5				
−5.5				
−6.5				
−7.5				
−8.5				

DISCUSSION

1. Based on the data of Table 16-1E and figure 16-2, what is the LSB voltage value for the value of R_F used in figure 16-1 for part 1(a)?

2. Repeat 1 for part 1(b).

3. For the value of R_F used in part 1(a) what is the MSB voltage?

4. The D/A converter of figure 16-1 has a four-step ladder. With the R_F of part 1(a) the resolution is poor. To improve the resolution, a fifth step can be added. What resistance value should be used in the ladder and what will the new value of the LSB be?

5. What will happen to the ladder resistance values as you attempt to make the value of the LSB smaller and smaller to give better resolution?

6. What effect does an incorrect resistor in the binary ladder have on the precision of the D/A conversion (see Fig. 16-4 and compare with Fig. 16-2a)? Discuss.

7. In part 2(c), explain why the pattern changes in steps.

CATHODE RAY OSCILLOSCOPE (CRO)*

CRO GAIN CALIBRATION AND POWER SUPPLY VOLTAGE CHECK

The basic measuring instrument used in these experiments is the CRO. It is used to observe and measure wave forms, but in particular it is used as a dc voltmeter. With proper care paid to the gain calibration of its vertical amplifiers, it can be used as an accurate laboratory measurement instrument. Since the ICs used in these experiments are fragile and have a very small voltage overload factor, the CRO can be and *must* be used to check power supply voltages *before* power is applied to the ICs.

Connect the output of the CRO internal calibrating voltage to the vertical input of the CRO. If a multirange calibrator is available, use a 10-volt p-p voltage and set the vertical gain to 2 volts/div. This will provide a p-p deflection of *exactly* 5 divisions. If 10 volts is not available, use a combination of calibrating voltage and vertical amplifier calibration to give a deflection of 4 or 5 divisions. Now obtain a steady pattern using the CRO triggering and horizontal TIME/DIV controls and check the voltage calibration of the vertical amplifier. If the deflection does not agree with the calibrating voltage, call the instructor.

IMPORTANT: Before connecting power to the ICs, check the power supply voltage against the calibrated CRO. If the measurement with the CRO agrees with the output voltage of the power supply, power can be applied to the ICs and the experiment can proceed. If it does not agree, call the instructor.

---

*It is assumed that the student has had some previous exposure to the CRO and is familiar with its basic operation and the function and operation of its controls.

Appendix B

LOGIC SYMBOLS

Logic functions are two-valued: ONE or ZERO, 1 or 0; HIGH or LOW, H or L; TRUE or FALSE, T or F; ON or OFF. While the functions can be implemented in many different physical ways, the net effect is that of identical logic performance. To show that identical logic is being performed, sets of graphic logic symbols have been developed to depict the various logic functions.

The logic symbols used in this manual are those based on the distinctive-shaped symbols of MIL STD 806B. These symbols are recommended by the semiconductor standardizing committees of the Electronics Industry Association (EIA) for registration of microelectronics logic functions. Virtually all integrated circuit (IC) manufacturers' published engineering data specifying dual-state logic functions use these symbols. Figure B-1 shows the symbols used in this manual.

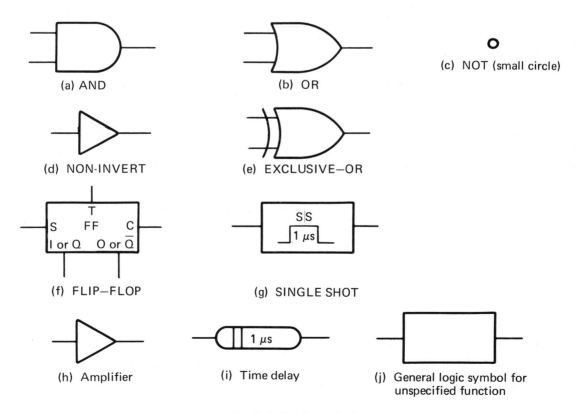

Fig. B-1. Logic symbols.

Logic functions may be simple or complex. A type of complexity that frequently occurs is the requirement that a gate be controlled by more inputs than can be provided by the number of available terminals of the case in which the gate is packaged. To provide additional gating capability, gates can be provided with an expander input. The symbol for this is shown in figure B-2.

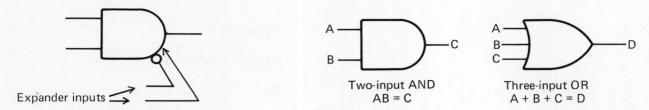

Fig. B-2. Expander input symbol.

Fig. B-3. Logic symbols for AND and OR functions.

Figure B-3 shows the application of these logic symbols to circuits such as those discussed in Chapter 1. In Chapter 1 the circuits were called AND and OR circuits. Integrated circuits which perform these functions are called AND and OR gates.

Although AND and OR circuits have many applications in logic circuitry, they do not have complete versatility. The NAND (AND plus inverter) and the NOR (OR plus inverter) have much greater usefulness. Logic symbols for NAND and NOR functions and INVERT are shown in figure B-4.

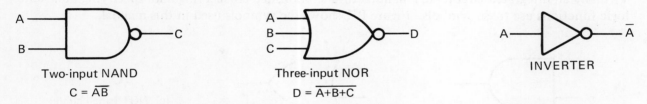

Fig. B-4. Logic symbols for NAND, NOR, and INVERT functions and applications.

Frequently, a function is controlled by a gate as part of a complex IC function. A typical illustration might be the control of the set input to a flip-flop by means of a three-input AND gate. Figure B-5 illustrates the logic-symbol method of illustrating this function.

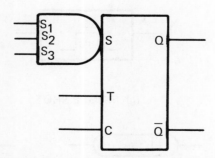

Fig. B-5. Flip-flop with SET function controlled by a three-input AND gate.

Appendix C

THE "UNIT LOAD" CONCEPT

To simplify output-fanout and input-fanin design, the UNIT LOAD (UL) concept has been developed. A UL has significance only for its own logic family and for compatible logic families, but not for noncompatible families.

A UL is the load presented by a standard gate input. For example, in the TTL gate of figure 4-1, a UL would be the current flowing out of input A when input A is at a low potential and input B is at a high potential. From figure 4-1 this is

$$\frac{(5 - 0.7)V}{4.0k\Omega} \approx 1.1 \text{ mA}$$

Now, let us suppose that output transistor Q_3 of figure 4-1 is capable of sinking 11 mA. We can express this on a UL basis as 10 UL.

UL values are shown on IC gate specification sheets as values in parentheses next to the IC package pin numbers. For example, suppose we looked at the logic diagram for the gate of figure 4-1 on a data sheet. It would be shown as in figure C–1.

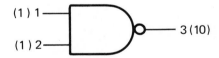

Fig . C-1. Logic diagram of the 2-input NAND gate of Fig. 4-1.

In figure C–1, the two gate inputs are at pins 1, 2, 4 and the output is at pin 3. The input loading, shown in parentheses, acts as 1 UL in TTL logic and the output can drive 10 UL (TTL unit loads) as shown in parentheses at the output terminal.

00

QUADRUPLE 2-INPUT POSITIVE-NAND GATES

positive logic:

$Y = \overline{AB}$

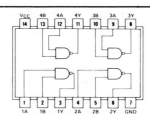

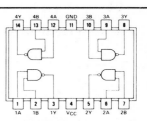

SN5400/SN7400(J, N)
SN54H00/SN74H00(J, N)
SN54L00/SN74L00(J, N)
SN54LS00/SN74LS00(J, N, W)
SN54S00/SN74S00(J, N, W)

SN5400/SN7400(W)
SN54H00/SN74H00(W)
SN54L00/SN74L00(T)

02

QUADRUPLE 2-INPUT POSITIVE-NOR GATES

positive logic:

$Y = \overline{A+B}$

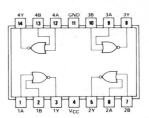

SN5402/SN7402(J, N)
SN54L02/SN74L02(J, N)
SN54LS02/SN74LS02(J, N, W)
SN54S02/SN74S02(J, N, W)

SN5402/SN7402(W)
SN54L02/SN74L02(T)

03

QUADRUPLE 2-INPUT POSITIVE-NAND GATES WITH OPEN-COLLECTOR OUTPUTS

positive logic:

$Y = \overline{AB}$

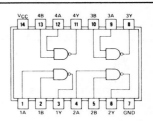

SN5403/SN7403(J, N)
SN54L03/SN74L03(J, N)
SN54LS03/SN74LS03(J, N, W)
SN54S03/SN74S03(J, N, W)

*Texas Instruments, Inc.

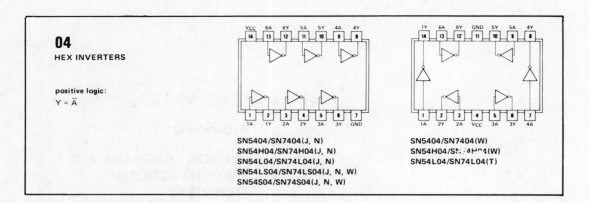

04
HEX INVERTERS

positive logic:
$Y = \overline{A}$

SN5404/SN7404(J, N)
SN54H04/SN74H04(J, N)
SN54L04/SN74L04(J, N)
SN54LS04/SN74LS04(J, N, W)
SN54S04/SN74S04(J, N, W)

SN5404/SN7404(W)
SN54H04/SN74H04(W)
SN54L04/SN74L04(T)

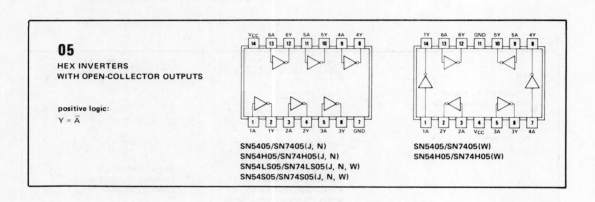

05
HEX INVERTERS
WITH OPEN-COLLECTOR OUTPUTS

positive logic:
$Y = \overline{A}$

SN5405/SN7405(J, N)
SN54H05/SN74H05(J, N)
SN54LS05/SN74LS05(J, N, W)
SN54S05/SN74S05(J, N, W)

SN5405/SN7405(W)
SN54H05/SN74H05(W)

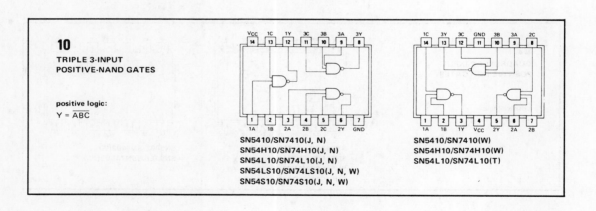

10
TRIPLE 3-INPUT
POSITIVE-NAND GATES

positive logic:
$Y = \overline{ABC}$

SN5410/SN7410(J, N)
SN54H10/SN74H10(J, N)
SN54L10/SN74L10(J, N)
SN54LS10/SN74LS10(J, N, W)
SN54S10/SN74S10(J, N, W)

SN5410/SN7410(W)
SN54H10/SN74H10(W)
SN54L10/SN74L10(T)

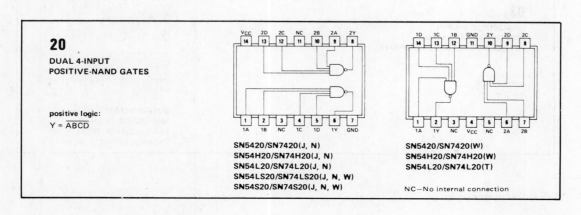

20
DUAL 4-INPUT
POSITIVE-NAND GATES

positive logic:
$Y = \overline{ABCD}$

SN5420/SN7420(J, N)
SN54H20/SN74H20(J, N)
SN54L20/SN74L20(J, N)
SN54LS20/SN74LS20(J, N, W)
SN54S20/SN74S20(J, N, W)

SN5420/SN7420(W)
SN54H20/SN74H20(W)
SN54L20/SN74L20(T)

NC—No internal connection

212

32

QUADRUPLE 2-INPUT
POSITIVE-OR GATES

positive logic:

$Y = A + B$

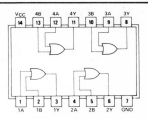

SN5432/SN7432(J, N, W)
SN54LS32/SN74LS32(J, N, W)

72

AND-GATED J-K MASTER-SLAVE FLIP-FLOPS WITH PRESET AND CLEAR

FUNCTION TABLE

INPUTS					OUTPUTS	
PRESET	CLEAR	CLOCK	J	K	Q	$\bar{Q}$
L	H	X	X	X	H	L
H	L	X	X	X	L	H
L	L	X	X	X	H*	H*
H	H	⎍	L	L	Q_0	$\bar{Q}_0$
H	H	⎍	H	L	H	L
H	H	⎍	L	H	L	H
H	H	⎍	H	H	TOGGLE	

positive logic: $J = J1 \cdot J2 \cdot J3$; $K1 \cdot K2 \cdot K3$

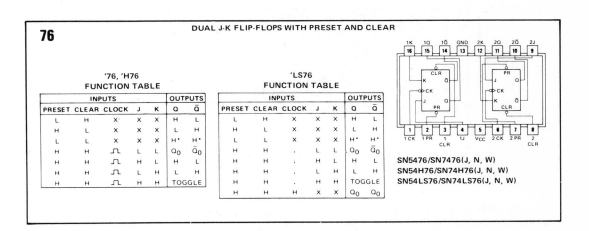

SN5472/SN7472(J, N)
SN54H72/SN74H72(J, N)
SN54L72/SN74L72(J, N)

SN5472/SN7472(W)
SN54H72/SN74H72(W)
SN54L72/SN74L72(T)

NC—No internal connection

76

DUAL J-K FLIP-FLOPS WITH PRESET AND CLEAR

'76, 'H76 FUNCTION TABLE

INPUTS					OUTPUTS	
PRESET	CLEAR	CLOCK	J	K	Q	$\bar{Q}$
L	H	X	X	X	H	L
H	L	X	X	X	L	H
L	L	X	X	X	H*	H*
H	H	⎍	L	L	Q_0	$\bar{Q}_0$
H	H	⎍	H	L	H	L
H	H	⎍	L	H	L	H
H	H	⎍	H	H	TOGGLE	

'LS76 FUNCTION TABLE

INPUTS					OUTPUTS	
PRESET	CLEAR	CLOCK	J	K	Q	$\bar{Q}$
L	H	X	X	X	H	L
H	L	X	X	X	L	H
L	L	X	X	X	H*	H*
H	H	.	L	L	Q_0	$\bar{Q}_0$
H	H	.	H	L	H	L
H	H	.	L	H	L	H
H	H	.	H	H	TOGGLE	
H	H	H	X	X	Q_0	Q_0

SN5476/SN7476(J, N, W)
SN54H76/SN74H76(J, N, W)
SN54LS76/SN74LS76(J, N, W)

TYPES SN5481A, SN5484A, SN7481A, SN7484A
16-BIT ACTIVE-ELEMENT MEMORIES

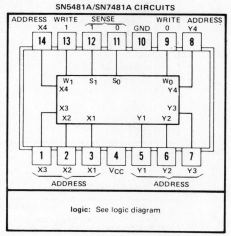

J OR N DUAL-IN-LINE
OR W FLAT PACKAGE (TOP VIEW)
SN5481A/SN7481A CIRCUITS

logic: See logic diagram

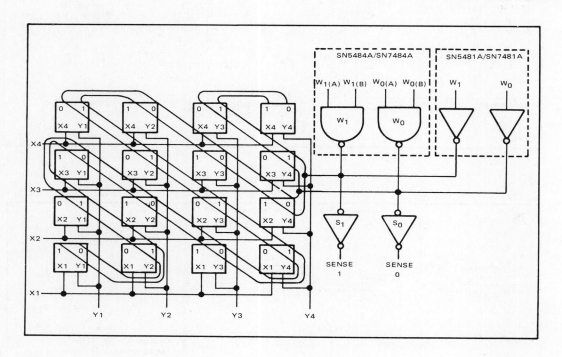

CIRCUIT TYPES SN5486, SN7486 QUADRUPLE 2-INPUT EXCLUSIVE-OR GATES

logic

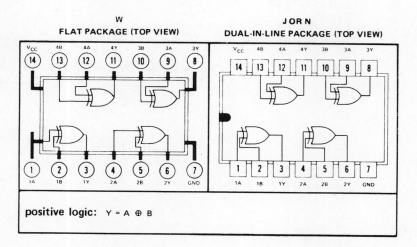

W FLAT PACKAGE (TOP VIEW)

J OR N DUAL-IN-LINE PACKAGE (TOP VIEW)

positive logic: $Y = A \oplus B$

TRUTH TABLE

INPUTS		OUTPUT
A	B	Y
0	0	0
0	1	1
1	0	1
1	1	0

CIRCUIT TYPES SN5490, SN7490
DECADE COUNTERS

J OR N
DUAL-IN-LINE PACKAGE (TOP VIEW)

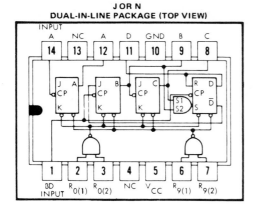

BCD COUNT SEQUENCE
(See Note 1)

COUNT	OUTPUT			
	D	C	B	A
0	0	0	0	0
1	0	0	0	1
2	0	0	1	0
3	0	0	1	1
4	0	1	0	0
5	0	1	0	1
6	0	1	1	0
7	0	1	1	1
8	1	0	0	0
9	1	0	0	1

RESET/COUNT (See Note 2)

RESET INPUTS				OUTPUT
$R_{0(1)}$	$R_{0(2)}$	$R_{9(1)}$	$R_{9(2)}$	D C B A
1	1	0	X	0 0 0 0
1	1	X	0	0 0 0 0
X	X	1	1	1 0 0 1
X	0	X	0	COUNT
0	X	0	X	COUNT
0	X	X	0	COUNT
X	0	0	X	COUNT

NC—No Internal Connection

NOTES: 1. Output A connected to input BD for BCD count.
2. X indicates that either a logical 1 or a logical 0 may be present.

TYPE SN74141
BCD-TO-DECIMAL DECODER/DRIVER

J OR N DUAL-IN-LINE
OR W FLAT PACKAGE (TOP VIEW)

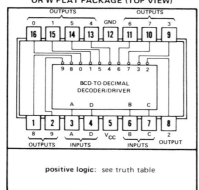

positive logic: see truth table

FUNCTION TABLE

INPUT				OUTPUT
D	C	B	A	ON[†]
L	L	L	L	0
L	L	L	H	1
L	L	H	L	2
L	L	H	H	3
L	H	L	L	4
L	H	L	H	5
L	H	H	L	6
L	H	H	H	7
H	L	L	L	8
H	L	L	H	9
H	L	H	L	NONE
H	L	H	H	NONE
H	H	L	L	NONE
H	H	L	H	NONE
H	H	H	L	NONE
H	H	H	H	NONE

H = high level, L = low level
[†]All other outputs are off

CIRCUIT TYPES SN52741, SN72741
OPERATIONAL AMPLIFIERS

J OR N DUAL-IN-LINE
PACKAGE (TOP VIEW)

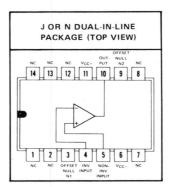

NC—No internal connection

215

ACCESS TIME. Time required to obtain information stored in a memory.

ACCUMULATOR. Section of a computer which performs the operation of addition and stores the sum.

A/D CONVERTER. Analog-to-digital converter. Circuitry which converts an analog voltage or current to an equivalent digital output.

ADDER. Circuitry which performs the operation of adding two numbers.

AMPLIFIER. Circuits which increase the amplitude and power level of an incoming signal.

ANALOG SWITCH. In D/A conversion, the circuitry which switches the binary information into the summing circuitry.

AND. Logic function whose output is a 1 when all inputs are equal to 1.

ASTABLE MULTIVIBRATOR. An oscillator having two outputs. The output levels are complementary; when one output is high the other is low.

ASYNCHRONOUS INPUT. Inputs to a flip-flop which affect the output of the flip-flop independent of all other inputs.

ASYNCHRONOUS OPERATION. An operation that proceeds independently within a system independent of any clock signals.

BCD. Abbreviation for BINARY CODED DECIMAL. 8421 code is commonly called BCD, although the 8421 code is sometimes called NBCD. All other codes are referred by their weighting; for example, 4221.

BINARY. See FLIP-FLOP.

BINARY CODED DECIMAL. A 4-bit binary code covering the decimal numbers 0, 1, 2, . . . , 9. Each digit requires a grouping of 4 bits.

BIPOLAR DEVICE. A device in which current flow occurs with both minority and majority carriers. npn and pnp transistors are examples of such devices.

BISTABLE. A circuit having two stable states.

BIT. A binary digit, either a 1 or a 0.

BOOLEAN ALGEBRA. The algebraic rules of logic.

BORROW. In subtraction, when the subtrahend is larger than the minuend, a borrow is required from the next more significant column.

BREAKDOWN VOLTAGE. The voltage at which a semiconductor device undergoes a sharp change in its current-voltage characteristic. There is a sharp increase in current at this voltage.

BYTE. A grouping of bits.

CARRY. In addition, when the sum exceeds the radix a carry is generated for addition in the next more significant column.

CLEAR. Set a binary to the 0 state.

CLOCK. In a digital system, a timing pulse for controlling the timing of operations.

COMPARATOR, ANALOG. A circuit which compares two analog voltages and indicates which is greater than or less than the other.

COMPARATOR, BINARY. An arrangement of logic gates for comparing two binary numbers for equality, less than, or more than. See also EXCLUSIVE-OR.

COMPLEMENT. (a) The sum of a number and its complement is equal to one less than a power of the radix. For example, in decimal the 9s complement of 57 equals 42. In binary the 1s complement of 1 is 0 and 0 is 1. Binary 1s complements are available from the opposite side of a FF. (b) The sum of a number and its complement is equal to a power of the radix. The 10s complement of 57 is equal to 43. In binary the 2s complement is obtained readily by adding one to the 1s complement. (c) The inverse of a logic function. $\bar{A}$ is the complement of A.

COMPLEMENTARY MOS. Also CMOS and COS/MOS. A series arrangement of n-channel and p-channel enhancement-mode transistors with the drains tied together. The output is taken from the common drain terminal.

COUNTER. Digital circuitry for counting incoming pulses and indicating the number of incoming pulses by the states of output terminal.

COUNTER, BINARY. Count states are represented in binary notation.

COUNTER, JOHNSON. See COUNTER, TWISTED RING.

COUNTER, MOD-N. Modified binary counter with N distinct count states. Returns to original state after N pulses are counted.

COUNTER, RING. A counter with output returned to input to form a continuous ring. One stage of the counter is in a different state from all other stages.

COUNTER, RIPPLE. A counter in which a count change progresses from one stage to the next so that a change ripples through the counter.

COUNTER, SYNCHRONOUS. A counter in which all stages change state at an identical time determined by a clock pulse.

COUNTER, TWISTED RING. A ring counter in which the state of the input stage is the complement of the final stage.

DIODE. Two-terminal device with high conduction in the forward voltage direction and essentially zero conduction in the reverse voltage direction.

DIVIDER (FREQUENCY). The output frequency is 1/N of the incoming frequency, where N is the number of different count states of the counter before returning to its initial state. See also COUNTER.

DOT-AND. Logic output of wired collector logic resulting in an AND function. Frequently called WIRED-OR.

DOT-OR. Logic output of wired collector logic resulting in an OR function.

DTL. Diode transistor logic.

EXCLUSIVE-OR. Digital circuit for two inputs that results in a 1 if one input is a 1 and the other is a 0. If both inputs are alike the output is a 0. It acts as a digital comparator, giving a 0 for like inputs and 1 for nonalike inputs.

FALSE. Equivalent of a 0 or an OFF.

FANIN. Number of inputs to a gate.

FLIP-FLOP (FF). A circuit having two stable states. Its state can be changed upon the application of a signal and will remain indefinitely in the new state after removal of the signal.

(a) R-S. Two inputs, R for reset and S for set. When the R input is activated the FF is placed into the 0 state. When the S input is activated the FF is placed into the 1 state. Activating both R and S simultaneously is not allowed.

(b) T. Upon application of a clock pulse to the T input the FF toggles.

(c) RST. Combination of R-S and T.

(d) J-K. Similar to RST, but clocked. J = R, K = S. In addition, when both J and K are activated the FF toggles. When J and K are not activated no change takes place when a clock pulse is applied.

(e) D (Data). Information is presented to the D input in digital form and appears at the output after a clock pulse.

FULL ADDER. Provides sum S and carry out C_o information for three inputs X, Y, and the carry in C_i information from the next lower significant column. Performs the operation $X + Y + C_i$.

FULL SUBTRACTOR. Provides difference D and borrow out B_0 information for three inputs X, Y, and the borrow in B_i information from the next lower significant column. Performs the operation $X - Y - B_i$.

GATE. Logic operator or circuit.

    (a) AND. Output is 1 if all inputs are 1.

    (b) NAND. AND followed by INVERT.

    (c) NOR. OR followed by INVERT.

    (d) OR. Output is 1 if any input is a 1.

HALF-ADDER. Performs the addition operation for the least significant bit and provides the sum S and carry out C_0 for the operation $X + Y$.

HALF-SUBTRACTOR. Performs the subtraction operation for the least significant bit and provides the difference D and borrow out B_0 for the operation $X - Y$.

HIGH LEVEL. More positive of the logic levels in a binary system.

INTEGRATED CIRCUIT (IC). An assembly of circuit components intended for operation as an electronic circuit. In a monolithic IC all components are fabricated simultaneously on a single silicon substrate. In a hybrid IC multiple substrates, individual semiconductors, and film components are combined.

INVERT. To perform the NOT function. Provides $\overline{A}$ when input A is applied.

ISOLATION DIODE. A p-n junction, normally reverse biased, which surrounds components in a monolithic IC to provide isolation from other components. Care must be taken in circuit operation to prevent the diodes from becoming forward biased.

KARNAUGH MAP. A map for logic equations arranged in a manner so that logic and circuit simplifications become readily apparent.

LADDER. In D/A conversion, resistor network for converting binary outputs on a weighted basis.

LATCH. A simple R-S flip-flop constructed with either two NAND or two NOR gates.

LOGIC DELAY. Speed of signal propagation through a gate. Frequently expressed as pair delay, which is the delay through a pair of inverting gates.

LOGIC DIAGRAM. A diagram using logic symbols to show the signal flow relationships in a system.

LOGIC SWING. Difference in voltage between the high voltage level and the low voltage level.

LOGIC SYMBOL. Graphic representation of basic logic function.

LOW LEVEL. More negative of the logic levels in a binary system.

LSB. Least Significant Bit.

LSI. Large-Scale Integration. A complex integrated circuit with more than 100 gates.

MASTER-SLAVE. A flip-flop with two separate clocked flip-flops. Information is first transferred into the master and then information from the master is transferred into the slave.

MEMORY. Information storage.

MONOLITHIC. See INTEGRATED CIRCUIT.

MONOSTABLE. Single shot, one shot. A circuit normally in a stable state. Upon application of a signal is triggered into a quasistable state and remains in the quasistable state for a time determined by an RC time constant. It then reverts to its stable state.

MOS. Metal Oxide Semiconductor. Field-effect transistor in which the gate is isolated from the channel by an oxide film.

MSI. Medium Scale Integration. A complex IC having between 12 and 100 gates.

NAND. AND followed by INVERT.

NEGATIVE LOGIC. The 1 level is more negative than the 0 level.

NOISE IMMUNITY—NOISE MARGIN. Difference between logic level and gate threshold voltage. Represents the maximum allowable spurious signal beyond which an undesired gate output level can occur.

NOR. OR followed by INVERT.

NOT. See INVERT.

OCTAL NUMBERS. Number system based upon radix 8.

ONE SHOT. See MONOSTABLE.

OPERATIONAL AMPLIFIER. High voltage gain, high input impedance, dc-wide band amplifier with output voltage level 0 volts when input voltage is zero. Single-input amplifier is inverting. With dual differential inputs, amplifier amplifies differential input.

PAIR DELAY. See LOGIC DELAY.

PARALLEL ADDER. An adder in which all bits are added simultaneously. See also SERIAL ADDER.

PARALLEL IN–SERIAL OUT. In a shift register, method of introducing parallel information into the register and retrieving information in serial form. See also SHIFT REGISTER and SERIAL IN–PARALLEL OUT.

PARASITIC ELEMENT. Undesired circuit components. Stray capacitances and lead inductances. In monolithic ICs, the isolation diodes.

PARITY. Error-checking method applied to a word which checks by counting the number of 1s. A parity bit is added to convert all words in the system to ODD parity or EVEN parity as required.

PASSIVE. A resistor is used between the output terminal and supply voltage.

POSITIVE LOGIC. The 1 level is more positive than the 0 level.

PRESET. See Set.

PULL-DOWN RESISTOR. A resistor connected to ground or to a negative voltage.

PULL-UP ACTIVE. A transistor is used between the output terminal and supply voltage to lower the output source impedance.

PULL-UP RESISTOR. A resistor connected to the positive supply voltage.

PULSE. A signal of short duration.

Q OUTPUT. Flip-flop output terminal corresponding to the TRUE output.

$\bar{Q}$ OUTPUT. Flip-flop output terminal corresponding to the FALSE output.

RACE. A timing conflict that occurs when two signals are coincidentally applied to a logic element.

RAM. Random Access Memory. A memory in which stored information can be retrieved within a time period independent of the location of the memory element within the memory. See also SCRATCH PAD MEMORY.

REGISTER. Storage location for binary information.

SHIFT. A register in which all information can be transferred one stage at a time by shift or clock pulses.

CIRCULATING SHIFT. A register in which the final stage is returned to the first stage so that information circulates within the register.

RESET. See CLEAR.

RING COUNTER. See COUNTER, RING.

RIPPLE. See COUNTER, RIPPLE.

RIPPLE COUNTER. See COUNTER, RIPPLE.

ROM. Read Only Memory. A memory in which information is permanently stored and cannot be altered.

RTL. Resistor Transistor Logic.

SCRATCH PAD MEMORY. A RAM for temporary information storage.

SERIAL ADDER. An adder in which each pair of bits is added one pair at a time. Requires less equipment than the PARALLEL ADDER but requires more time to perform the operation.

SERIAL IN–PARALLEL OUT. In a shift register, introduction of information into the register in serial form and retrieval of information in parallel form. See also PARALLEL IN–SERIAL OUT.

SET. In a flip-flop, to set into the 1 state. Opposite of RESET.

SHIFT. In a register, to transfer all information by one stage upon application of a clock or shift pulse.

SHIFT REGISTER. See REGISTER, SHIFT.

SSI. Small-Scale Integration. An IC with less than 10 gates.

STROBE. To clock or gate.

SYNCHRONOUS. A digital system in which all operations occur simultaneously at a time determined by a clock pulse.

SYNCHRONOUS INPUTS. Inputs to a flip-flop which control the output but only after the application of a clock pulse. For example, J-K inputs in J-K flip-flop.

THRESHOLD VOLTAGE. Gate voltage at which the output level transfers from HIGH LEVEL to LOW LEVEL.

TRUTH TABLE. Tabulation of output level for all possible valid input combinations.

TTL, T^{2}L. Transistor-Transistor Logic.

TRUE. ONE state. Opposite of FALSE.

TURN-OFF TIME. Delay time after pulse initiation to cut off current in a transistor.

TURN-ON TIME. Delay time after pulse initiation for transistor current conduction.

UP–DOWN COUNTER. Capable of being reversible with respect to its count state procedure.

WIRED COLLECTOR LOGIC. External connection of outputs of logic gates, to provide additional order of logic gating. Also called WIRED-OR, PHANTOM-OR, COLLECTOR-OR. See also DOT-AND and DOT-OR.

WORD. A grouping of bits treated as an entity in a computer.

ZENER DIODE. Diode with operation in the current region beyond reverse voltage breakdown.

ZERO. Logic level opposite to ONE. Equivalent to FALSE.

Appendix F

REFERENCES

TEXTBOOKS

Burr-Brown, *Operational Amplifiers*. McGraw-Hill, New York, 1971.

Fitchen, F. C., *Electronic Integrated Circuits and Systems*, Van Nostrand Reinhold, New York, 1970.

Kintner, Paul M., *Electronic Digital Techniques*, McGraw-Hill, New York, 1968.

Maley, Gerald A., *Manual of Logic Circuits*, Prentice-Hall, Englewood Cliffs, N.J., 1970.

Malvino, Albert P., and Leach, Donald P., *Digital Principles and Applications,* McGraw-Hill, New York, 1969.

Mandl, Mathew, *Electronic Switching Circuits, Boolean Algebra and Mapping*, Prentice-Hall, Englewood Cliffs, N.J., 1969.

Mandl, Mathew, *Fundamentals of Electronic Computers, Digital and Analog*, Prentice-Hall, Englewood Cliffs, N.J., 1967.

Nashelsky, Louis, *Digital Computer Theory*, Wiley, New York, 1966.

Wickes, William E., *Logic Design with Integrated Circuits*, Wiley, New York, 1968.

TECHNICAL PUBLICATIONS

Fairchild Semiconductor:
   APP 107 Diode Transistor Micrologic.
   Linear Integrated Circuits, Applications Handbook.

Motorola Semiconductor Products, Inc.:
   AN-235 Using the Motorola MDTL Line of Integrated Circuits.
   AN-262A Decade Counters Using MDTL Integrated Circuits.
   AN-284 MDTL IC Shift Registers.
   AN-409 MDTL Multivibrator Circuits.
   AN-464 MTTL Designers Note—The MC4004/MC4005. A 16-Bit RAndom Access Memory.

National Semiconductor:
   AN-20 Applications Guide for Operational Amplifiers.

Texas Instruments:
   CA-102 TTL Integrated Circuits: Counters and Registers.
   CA-112 Logic Design with Series 54/74 Gates.

TECHNICAL ARTICLES

Garrett, Lane S., Integrated-Circuit Digital Logic Families.
   IEEE Spectrum, Oct. 1970, Part 1, pp 46-58.

IEEE Spectrum, Nov. 1970, Part 2, pp 63-72.
IEEE Spectrum, Dec. 1970, Part 3, pp 30-42.
A comprehensive summary and comparison of the digital logic families.

This appendix gives the construction details of a laboratory IC socket and switch bank for use in the experiments of the manual. The socket is a 16-pin dual in-line socket and is used with both the 14-pin and 16-pin ICs used for the experiments. All the experiments can be performed with a maximum of six sockets and two switch banks.

To minimize damage to the IC pins when extracting the ICs from the socket, use the AUGAT dual in-line extractor tool T114-1.

The POMONA banana plug patch cord series Model B mates with the five-way binding posts and are stackable. The 8″ length is the most useful for intersocket connections and to discrete components. The 4″ length has limited usefulness and can be used only between adjacent pins. The 12″ length is needed for the more complex set-ups. For connections to power supplies, the CRO, signal sources, and other equipment, lengths up to 60″ are available. Model HB is an extension stacking series that can be used for extension to longer length. Pomona also has available five-way binding posts to BNC and double binding post to BNC adapters.

The CRO used for the experiments should be a triggered sweep CRO. A dual-beam or dual-trace CRO, while not essential, can be helpful in the analysis of sequential patterns.

Many of the experiments require a pulse generator or square wave generator (SWG) with positive-going output and voltage capabilities up to +5 volts. The frequency range is from 1 Hz to approximately 50 kHz. Single-pulse capability is essential. The output impedance can be between 50 to 500 Ω. If the generator output goes both positive and negative, the negative portion can be clipped with a small-signal diode. Care must be taken to ensure that the positive-going voltage applied to the IC does not exceed V_{CC} at any time.

Several experiments require an audio oscillator.

Power supply requirements can be found in the individual experiments. The supplies should be well regulated and have low RF output impedance to minimize interaction and noise generation.

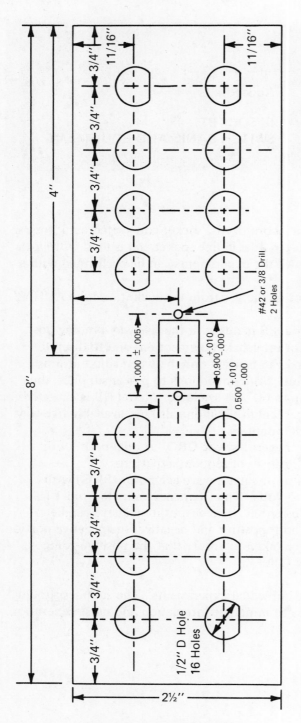

11/16''
11/16''
3/4''
3/4''
3/4''
3/4''
4''
#42 or 3/8 Drill
2 Holes
1.000 ± .005
0.900 +.010 −.000
0.500 +.010 −.000
8''
3/4''
3/4''
3/4''
3/4''
1/2'' D Hole
16 Holes
2½''

Note — Mount socket below.

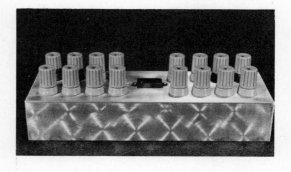

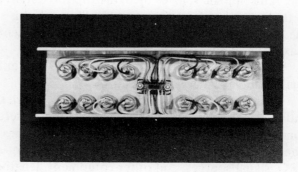

PARTS LIST
1   IC Socket Augat No. 316-AG3A
    Jermyn No. A1229AM
16 5-Way Binding Posts
2   RH # 2 Machine Screws x½''
2   # 2 Hex Nuts
2   # 2 Lock Washers
1   Aluminum channel 8''x2½''x1½''x
    1/8'', # 6063T5 or equivalent
    Peter Frassee, Inc.
    17 Grand St., New York, N.Y. 10013